Financial Schemes for Resilient Flood Recovery

Financial schemes for flood recovery, if properly designed and implemented, might increase flood resilience. However, options for the increase of flood resilience during the recovery phase are to a large extent overlooked and the diversity of existing schemes shows that there has been a lack of consensus on how to achieve resilient flood recovery.

Financial Schemes for Resilient Flood Recovery investigates how the implementation of financial schemes (government relief subsidies, insurance schemes, buy-outs, etc.) might increase flood resilience. The chapters included in this edited volume address the following questions: Shall government relief subsidies exist when there is flood insurance in place, and, if so, how might they both be coordinated? Where (or how) to decide about build back better incentives and where to go for planned relocation programs? What is the distributional equity of financial schemes for flood recovery, and has it been sufficiently treated?

The book covers different approaches to flood recovery schemes with specific intervention rationales in different countries. Empirical evidence provided clearly shows the great diversity of financial flood recovery schemes. This diversity of state-funded schemes, private-based insurance schemes, and hybrids as well as planned relocation schemes indicates a lack of a consistent and strategic approach in flood risk management and flood resilience about flood recovery.

The chapters in this book were originally published in the *Environmental Hazards*.

Lenka Slavíková is Associate Professor in public economics at J.E. Purkyně University, Ústí nad Labem, Czech Republic. She specializes in water and biodiversity governance with the focus on Central and Eastern European Countries. She investigates flood risk perception of household and municipalities and financial instruments to achieve flood resilience.

Thomas Hartmann is Associate Professor at Wageningen University, the Netherlands, and he teaches at J.E. Purkyně. University, Ústí and Laben, Czech Republic, and Bonn University, Germany. He combines an engineering perspective with socio-political approaches to flood risk management and land policies and has published numerous papers, books, and special issues on these topics.

Thomas Thaler is Research Fellow at the Institute of Mountain Risk Engineering, University of Natural Resources and Life Sciences, Vienna, Austria. He focuses on design and effectiveness of natural hazard governance systems as well as integrating European environmental policies into national and local institutions.

Financial Schemes for Resilient Flood Recovery

Edited by
Lenka Slavíková, Thomas Hartmann and Thomas Thaler

Routledge
Taylor & Francis Group
LONDON AND NEW YORK

First published 2022
by Routledge
2 Park Square, Milton Park, Abingdon, Oxon, OX14 4RN

and by Routledge
605 Third Avenue, New York, NY 10158

Routledge is an imprint of the Taylor & Francis Group, an informa business

British Library Cataloguing-in-Publication Data
A catalogue record for this book is available from the British Library

ISBN13: 978-1-032-01756-3 (hbk)
ISBN13: 978-1-032-01758-7 (pbk)
ISBN13: 978-1-003-17986-3 (ebk)

Typeset in Myriad Pro
by codeMantra

Publisher's Note
The publisher accepts responsibility for any inconsistencies that may have arisen during the conversion of this book from journal articles to book chapters, namely the inclusion of journal terminology.

The Introduction and Chapters 1-5 in this book were based upon work from COST Action LAND4FLOOD Nature Flood Retention on Private Land, supported by COST (European Cooperation in Science and Technology), www.cost.eu.

Contents

Citation Information

The chapters in this book were originally published in different issues of the *Environmental Hazards*. When citing this material, please use the original citations and page numbering for each article, as follows:

Introduction

Financial schemes for resilient flood recovery
Lenka Slavíková, Thomas Hartmann and Thomas Thaler
Environmental Hazards, volume 19, issue 3 (March 2020) pp. 223–227

Chapter 1

Measuring social equity in flood recovery funding
Christopher T. Emrich, Eric Tate, Sarah E. Larson and Yao Zhou
Environmental Hazards, volume 19, issue 3 (March 2020) pp. 228–250

Chapter 2

Approaches to state flood recovery funding in Visegrad Group Countries
Lenka Slavíková, Pavel Raška, Kazimierz Banasik, Marton Barta, Andras Kis, Silvia Kohnová, Piotr Matczak and Ján Szolgay
Environmental Hazards, volume 19, issue 3 (March 2020) pp. 251–267

Chapter 3

Financial recovery schemes in Austria: how planned relocation is used as an answer to future flood events
Thomas Thaler and Sven Fuchs
Environmental Hazards, volume 19, issue 3 (March 2020) pp. 268–284

Chapter 4

The French Cat' Nat' *system: post-flood recovery and resilience issues*
Bernard Barraqué and Annabelle Moatty
Environmental Hazards, volume 19, issue 3 (March 2020) pp. 285–300

Chapter 5

An assessment of best practices of extreme weather insurance and directions for a more resilient society
P. Hudson, L.T. De Ruig, M.C. de Ruiter, O.J. Kuik, W.J.W. Botzen, X. Le Den, M. Persson, A. Benoist and C.N. Nielsen
Environmental Hazards, volume 19, issue 3 (March 2020) pp. 301–321

Chapter 6

Disaster, relocation, and resilience: recovery and adaptation of Karamemedesane in Lily Tribal Community after Typhoon Morakot, Taiwan, Environmental Hazards
Sasala Taiban, Hui-Nien Lin and Chun-Chieh Ko
Environmental Hazards, volume 19, issue 2 (2020) pp. 209–222

Chapter 7

Post-disaster communalism: land use, ownership, and the shifting 'publicness' of urban space in recovery
Elyse M. Zavar and Ronald L. Schumann III
Environmental Hazards, volume 19, issue 4 (2020) pp. 398–416

Chapter 8

Prospects for disaster management in China and the role of insurance
Xian Xu
Environmental Hazards, volume 18, issue 5 (2019) pp. 383–399

Contributors

Kazimierz Banasik Faculty of Civil and Environmental Engineering, Department of River Engineering, Warsaw University of Life Sciences – SGGW, Poland.

Bernard Barraqué Centre International de Recherches sur l'Environnement et le Développement, Paris, France.

Marton Barta Regional Centre for Energy Policy Research, Corvinus University of Budapest, Hungary.

A. Benoist Ramboll Consulting, Copenhagen S, Denmark.

W.J.W. Botzen Institute for Environmental Studies, VU University, Amsterdam, The Netherlands. Utrecht University School of Economics (U.S.E.), Utrecht University, The Netherlands. The Wharton School, Risk Management and Decision Processes Center, University of Pennsylvania, Philadelphia, USA.

L.T. De Ruig Institute for Environmental Studies, VU University, Amsterdam, The Netherlands.

M.C. de Ruiter Institute for Environmental Studies, VU University, Amsterdam, The Netherlands.

Christopher T. Emrich School of Public Administration, National Center for Integrated Coastal Hazards, and Sustainable Coastal Systems Cluster, University of Central Florida, Orlando, USA.

Sven Fuchs Institute of Mountain Risk Engineering, University of Natural Resources and Life Sciences, Vienna, Austria.

Thomas Hartmann Department of Environmental Sciences, Wageningen Univeristy, The Netherlands.

P. Hudson Institute of Environmental Science and Geography, University of Potsdam, Germany.

Andras Kis Regional Centre for Energy Policy Research, Corvinus University of Budapest, Hungary.

Chun-Chieh Ko Assembly of Ngudradrekai, Pingtung, Taiwan.

Silvia Kohnová Faculty of Civil Engineering, Department of Land and Water Resources Management, Slovak University of Technology, Bratislava, Slovakia.

O.J. Kuik Institute for Environmental Studies, VU University, Amsterdam, The Netherlands.

Sarah E. Larson School of Public Administration, University of Central Florida, Orlando, USA.

X. Le Den Ramboll Consulting, Copenhagen S, Denmark.

Hui-Nien Lin Indigenous Program of the College of Tourism and Hospitality, I-Shou University, Kaohsiung, Taiwan.

Piotr Matczak Institute of Sociology, Adam Mickiewicz University, Poznań, Poland.

Annabelle Moatty Laboratoire de Géographie Physique, (UMR 8591) CNRS, Université Paris I Panthéon-Sorbonne, UPEC, Meudon, France.

C.N. Nielsen Ramboll Consulting, Copenhagen S, Denmark.

M. Persson Ramboll Consulting, Copenhagen S, Denmark.

Pavel Raška Faculty of Science, J. E. P. University in Usti nad Labem, Usti nad Labem, Czech Republic.

Ronald L. Schumann III Department of Emergency Management & Disaster Science, University of North Texas, Denton, USA.

Lenka Slavíková Faculty of Social and Economic Studies, J.E.Purkyně University in Ústí nad Labem, Usti nad Labem, Czech Republic.

Ján Szolgay Faculty of Civil Engineering, Department of Land and Water Resources Management, Slovak. University of Technology, Bratislava, Slovakia.

Sasala Taiban Department of Sociology, National Sun Yat-sen University, Kaohsiung, Taiwan.

Eric Tate Department of Geographical & Sustainability Sciences, University of Iowa, Iowa City, USA.

Thomas Thaler Institute of Mountain Risk Engineering (IAN), University of Natural Resources and Life Sciences, Vienna, Austria.

Xian Xu School of Economics, Fudan University, Shanghai, People's Republic of China.

Elyse M. Zavar Department of Emergency Management & Disaster Science, University of North Texas, Denton, USA.

Yao Zhou School of Public Administration, National Center for Integrated Coastal Hazards, and Sustainable Coastal Systems Cluster, University of Central Florida, Orlando, USA.

Introduction: financial schemes for resilient flood recovery

Lenka Slavíková, Thomas Hartmann and Thomas Thaler

ABSTRACT
Options for the increase of flood resilience during the recovery phase are, to a large extent, overlooked. The special issue **Financial Schemes for Resilient Flood Recovery** investigates how the implementation of financial schemes (government relief subsidies, insurance schemes, buy-outs, etc.) might increase flood resilience. Papers address the following questions: Shall government relief subsidies exist when there is flood insurance in place, and, if so, how might they both be coordinated? Where (or how) to decide about build back better incentives and where to go for planned relocation programs? What is the distributional equity of financial schemes for flood recovery, and has it been sufficiently treated?

In Europe, large fluvial floods are currently considered the most expensive natural hazard event (Munich Re, 2018). Similarly in the U.S., numerous areas have experienced repeated major flooding due to changing rainfall patterns (Livingston, 2019). According to the recent report of the Intergovernmental Panel for Climate Change, these trends might even be accelerated after 2030 if climate-change mitigation efforts fail (IPCC, 2018). There is also a high degree of spatial and temporal variability and uncertainty regarding particular impacts of watershed ecosystem services, including flood control (Tsvetkova & Randhir, 2019). The need for flood damage reimbursement mechanisms, especially after large flooding, is widely recognized. Their aim is to return society to "normal" in a timely and efficient manner (Liao et al., 2012). But does "normal" mean the same as before the flood? And if not, who ought to incentivize affected people to build back "better" and how can this be accomplished?

The concept of flood resilience is increasingly discussed in academia in this context (Fekete, Hartmann, & Jüpner, 2019). According to Schelfaut et al. (2011, p. 826), flood-resilient communities are able "to reduce, prevent and cope with the flood risks"... and they "have improved their capacity in each phase of the flood management cycle", including the recovery phase. However, what is meant by resilience depends on the discipline and specific perspective. The academic debate even considers resilience as a "boundary object" (Baggio, Brown, & Hellebrandt, 2015; Fekete et al., 2019), which means a concept that derives intrinsic value from its fuzziness and thereby allows different disciplines to communicate with each other (Brand & Jax, 2007). Etymologically, resilience at its core can be translated as "bouncing-back" (Alexander, 2013; Rodina, 2018), meaning it describes the ability of a system to absorb shock events (Holling, 1973). The debate on resilience embraces the ability to improve the shock-absorbing capacity of a system as well (Munich Re, 2016; Raška, Stehlíková, Rybová, & Aubrechtová, 2019). This special issue focuses on the role of financial flood recovery schemes to not only bounce back but also to bounce forward. Options for the increase of flood resilience during the recovery phase are, to a large extent, overlooked (Sandink, Kovacs, Oulahen, & Shrubsole, 2016).

As a result, recovery funding often aims solely at early restoration (Thomalla et al., 2018). Flood victims are regarded as mere recipients of money, but not as key stakeholders in preventing damages during the next event (Boustan, Kahn, & Rhode, 2012; Seifert, Botzen, Kreibich, & Aerts, 2013). The recent evidence shows that numerous flood recovery schemes have not been used strategically to increase resilience (Priest, Penning-Rowsell, & Suykens, 2016; Slavíková, 2016). Contrarily, subsidized recovery has generally maintained or even increased the vulnerability of affected communities (Sandink et al., 2016). Only a few authors have addressed the factual difficulty of reaching the recovery-mitigation nexus through the rigorous analysis of existing financial flood recovery schemes (Suykens, Priest, van Door-Hoekveld, Thuillier, & van Rijswick, 2016). So far no consensus has been reached upon how to achieve efficient and/or successful recovery.

The special issue **Financial Schemes for Resilient Flood Recovery** addresses the solidarity versus the effectiveness of flood recovery financing. It investigates how the implementation of financial schemes (government relief subsidies, insurance schemes, buy-outs, etc.) might increase flood resilience within existing institutional contexts. In particular, the following questions are covered: Shall government relief subsidies exist when there is flood insurance in place, and, if so, how might they both be coordinated? Where (or how) to decide about build back better incentives and where to go for planned relocation programs? What is the distributional equity of financial schemes for flood recovery, and has it been sufficiently treated? The transnational knowledge transfer regarding impacts of different schemes in different institutional contexts is needed to enhance learning and produce synergies. It may result in better integration of the recovery phase into overall flood damage mitigation efforts.

The special issue consists of eight papers covering different approaches to flood recovery schemes with specific intervention rationales in different countries. It opens up with two papers devoted to design, role, and distributional aspects of government relief subsidies (Emrich, Tate, Larson, & Zhou, 2019; Slavíková et al., 2019). Thaler and Fuchs (2019) discuss the reverse links between government recovery subsidies and the subsequent funding schemes for planned relocation. Zavar and Schumann III (2020) deal with the post-disaster emergence of public-private hybrid spaces when buy-outs are one of the key factors of this evolution. Taiban et al. (2020) introduce the impacts of forced relocation on the resilience of indigenous communities. The contribution by Barraqué & Moatty (2020) redirects the reader's attention to the functioning of insurance schemes guaranteed by the state and shows how the resilience component is difficult to incorporate into existing compensation rules. Xu (2019) provides an overview of financial recovery schemes in China, including commercial insurance, disaster insurance with government participation, and their mutual linkages. Finally, numerous private-based disaster insurance schemes are tested for their resilience impacts, and good practices are recommended (Hudson et al., 2019). These papers cover schemes that are mainly focusing on just recovery to schemes that try to trigger adaptation (even to the level of relocation). Also, the contributions show the different financial models – reaching from state-financed recovery schemes to insurance-based schemes. So, each paper contributes in specific ways to an inventory of different schemes and evaluates and discusses them critically:

In *Measuring social equity in flood recovery funding* Emrich et al. (2019) focus on social inequities as barriers to access post-disaster compensations. They refer to social dimensions of resilience and investigate the distributive equity in four flood recovery funding programs in the United States. Within the statistical analysis, the linkages among the amount of compensation received and the respective levels of damage, income, housing tenure, age, etc., were tested. The results contribute to a rare knowledge base of how social characteristics influence the extent of recovery support and what kinds of adjustments of the existing programs might better address the inequity problem. The contribution on *Approaches to state flood recovery funding in Visegrad group countries* teases out similarities and differences of government relief funding schemes in four postsocialistic European countries: Czech Republic, Hungary, Poland, and Slovakia (Slavíková et al., 2019). It seeks to answer if (and to what extent) the resilience incentives are incorporated within recovery funding. The analysis also highlights specifics in the development of flood risk-sharing strategies in the environment of inhabitants' strong historical reliance on the state.

The problem of insufficient policy coordination is raised by the contribution of *Financial recovery schemes in Austria: How planned relocation is used as an answer to future flood events* (Thaler & Fuchs, 2019). With the use of expert interviews, authors search for linkages between government relief funding and voluntary planned relocation programs in Austria. Their intention is to shed light on timing, incentives, and factors that influence people's willingness (or lack thereof) to sell out recently renovated houses. In contrast, Taiban et al. (2020) in *Disaster, relocation, and resilience: recovery and adaptation of Karamemedesane in Lily Tribal Community after Typhoon Morakot, Taiwan* brink the evidence on forced allocation impacts on the resilience of indigenous communities in Taiwan. They highlight the linkage between disaster risk reduction and social vulnerability increase during relocation efforts. They also argue that the connection of local people to the land, land-based practices, and culture are crucial factors of resilient relocation and recovery. Finally, relocation programs (buy-outs) are investigated by Zavar and Schumann III (2020) in *Post-disaster communalism: land use, ownership, and the shifting 'publicness' of urban space in recovery* as one cause of the emergence of public-private hybrid spaces. Building on field data, surveys, and interviews from two U. S. sites, authors conclude that communal land-use practices in urban areas may increase enabling more flexible disaster recovery.

Barraqué & Moatty (2020), in their essay *The French Cat'Nat' system: Post-flood recovery and resilience issues*, investigate the potential for build-back better incentives within the specific government-guaranteed insurance-based scheme in France. They contrast the relative success of this scheme (speed of compensation, broad coverage) with its low sustainability (repeated compensations for the same properties) and the moral-hazard behaviour of French municipalities responsible for the development reduction in floodplains. The paper also suggests a few risk reduction improvements for existing relief funding and land-use planning tools. The issue of efficiency versus availability of different financial recovery schemes is raised also by Xu (2019) in *Prospects for disaster management in China and the role of insurance*. He calls for the enhancement of the role of commercial insurance and re-structuring of governmental aid programs that may disincentive people from more resilient behaviour.

Finally, the paper *An assessment of best practices of extreme weather insurance and directions for a more resilient society* (Hudson et al., 2019) uses multicriteria analysis to evaluate the resilience components in private property and agricultural insurance schemes in 12 different European countries. The focus is not just on floods but all different types of extreme weather situations. Based on the evaluation results, the authors recommend best practices for insurance schemes themselves as well as for their coordination with other recovery mechanisms (such as having the state compensations available only to those who bought insurance before the hazard event).

Empirical evidence provided in these contributions to the special issue clearly shows the great diversity of financial flood recovery schemes. This diversity of state-funded schemes, private-based insurance schemes, and hybrids (Hudson et al., 2019; Slavíková et al., 2019) as well as planned relocation schemes (Thaler & Fuchs, 2019) indicates a lack of a consistent and strategic approach in flood risk management and flood resilience with regard to flood recovery. This might call for more strategic use of these schemes in the future providing their multiple impacts (Taiban et al., 2020; Zavar & Schumann III, 2020). Also, the coordination of other flood risk management policies with recovery funding is at best in its infancy, though the contributions indicate that there is a big potential to re-think such schemes to avoid increasing or repeating damages (respectively avoid moral hazards in the first place) (Xu, 2019), for example by connecting land-use restrictions with compensation rules (Barraqué & Moatty, 2020). The analysis of the problem of equal access to flood damage compensations and their just distributions (Emrich et al., 2019) reveals multiple challenges for aiming to contribute to flood resilience in the recovery phase. In particular, it is worth further discussing if focus on resilience increases the inequity of financial schemes and vice versa.

Aknowledgements

This introduction as well as the entire special issue *Financial Schemes for Resilient Flood Recovery* was compiled within the COST Action LAND4FLOOD Nature Flood Retention on Private Land, supported by COST (European Cooperation in Science and Technology). COST (cost.eu) is a funding agency for research and innovation networks. Our actions help to connect research initiatives across Europe and enable scientists to grow their ideas by sharing them with their peers. This boosts their research, careers and innovation records.

References

Alexander, D. E. (2013). Resilience and disaster risk reduction. An etymological journey. *Natural Hazards and Earth System Sciences, 13*(11), 2707–2716. doi:10.5194/nhess-13-2707-2013

Baggio, J. A., Brown, K., & Hellebrandt, D. (2015). Boundary object or bridging concept? A citation network analysis of resilience. *Ecology & Society, 20*(2). doi:10.5751/ES-07484-200202

Barraqué, B. & Moatty, A. (2020). The French *Cat'Nat'* system: Post-flood recovery and resilience issues. *Environmental Hazards, 19*(3), 285–300. https://www.tandfonline.com/doi/abs/10.1080/17477891.2019.1696738?journalCode=tenh20

Boustan, L. P., Kahn, M. E., & Rhode, P. W. (2012). Moving to higher ground: Migration response to natural disasters in the early twentieth century. *Americal Economic Reiview, 102*(3), 238–244.

Brand, F. S., & Jax, K. (2007). Focusing the meaning(s) of resilience: Resilience as a descriptive concept and a boundary object. *Ecology & Society, 12*(1), 23. Retrieved from http://www.ecologyandsociety.org/vol12/iss1/art23/

Emrich, C., Tate, E., Larson, S., & Zhou, Y. (2019). Measuring social equity in flood recovery funding. *Environmental Hazards*. doi:10.1080/17477891.2019.1675578

Fekete, A., Hartmann, T., & Jüpner, R. (2019). Resilience: On-going wave or subsiding trend in flood risk research and practice? WIRES Water. doi:10.1002/wat2.1397

Holling, C. S. (1973). Resilience and stability of ecological systems. *Annual Review of Ecology and Systematics, 4*(1), 1–23. doi:10.1146/annurev.es.04.110173.000245

Hudson, P., De Ruig, L. T., de Ruiter, M. C., Kuik, O. J., Botzen, W. J. W., Le Den, X., … Nielsen, C. N. (2019). An assessment of best practices of extreme weather insurance and directions for a more resilient society. *Environmental Hazards*. doi:10.1080/17477891.2019.1608148

IPCC (2018). Global warming of 1.5°C. The intergovernmental panel for climate change. On-line (10 September 2019). Retrieved from https://www.ipcc.ch/sr15/download/

Liao, K.-H. (2012). A theory on urban resilience to floods—A basis for alternativeplanning practices. *Ecology & Society, 17*(4), 48.

Livingston, I. (2019). A 125-year weather record just broke in the US, but the floods will keep coming. Washington Post. On-line (5th October 2019). Retrieved from https://www.sciencealert.com/the-water-just-keeps-coming-as-the-us-continues-its-wettest-12-months-in-125-years

Munich Re (2016). Naturkatastrophen 2016 - Analysen, Bewertungen, Positionen. Land Unter. TOPIC GEO. Munich.

Munich Re (2018). Anniversary of historic floods in Germany: What is the best protection against flooding? On-line (7 October 2019). Retrieved from https://www.munichre.com/topics-online/en/climate-change-and-natural-disasters/natural-disasters/floods/flood-europe-2018.html

Priest, J. S., Penning-Rowsell, E. C., & Suykens, C. (2016). Promoting adaptive flood risk management: The role and potential of flood recovery mechanisms. *E3S Web of Conferences* 7/17005.

Raška, P., Stehlíková, M., Rybová, K., & Aubrechtová, T. (2019). Managing flood risk in shrinking cities: Dilemmas for urban development from the Central European perspective. *Water International, 44*. doi:10.1080/02508060.2019.1640955

Rodina, L. (2018). Defining "water resilience". Debates, concepts, approaches, and gaps. *WIREs Water, 6*(2), e1334. doi:10.1002/wat2.1334

Sandink, D., Kovacs, P., Oulahen, G., & Shrubsole, D. (2016). Public relief and insurance for residential flood losses in Canada: Current status and commentary. *Canadian Water Resources Journal, 41*(3/4), 220. doi:10.1080/07011784.2015.1040458

Seifert, I., Botzen, W. J. W., Kreibich, H., & Aerts, J. C. J. H. (2013). Influence of flood risk characteristics on flood insurance demand: A comparison between Germany and the Netherlands. Natural Hazards and Earth System Sciences, *13*, 1691–1705.

Schelfaut, K., Pannemans, B., van der Craats, I., Krywkow, J., Mysiak, J., & Cools, J. (2011). Bringing flood resilience into practice: The FREEMAN project. *Environmental Science and Policy, 14*(7), 825–833. doi:10.1016/j.envsci.2011.02.009

Slavíková, L. (2016). Effects of government flood expenditures: The problem of crowding-out. *Journal of Flood Risk Management, 11*(1), 95.

Slavíková, L., Raška, P., Banasik, K., Barta, M., Kis, A., Kohnova, S., … Szolgay, J. (2019). Approaches to state flood recovery funding in Visegrad Group Countries. *Environmental Hazards*. doi:10.1080/17477891.2019.1667749

Suykens, C., Priest, S. J., van Door-Hoekveld, W. J., Thuillier, T., & van Rijswick, M. (2016). Dealing with flood damages: Will prevention, mitigation and ex post compensation provide for a resilient triangle? *Ecology & Society, 21*(4), 1.

Taiban, S., Hui-Nien, L., & Chun-Chieh, K. (2020). Disaster, relocation, and resilience: Recovery and adaptation of Karamemedesane in Lily Tribal Community after Typhoon Morakot, Taiwan. *Environmental Hazards, 19*(2), 209–222. doi:10.1080/17477891.2019.1708234

Thaler, T., & Fuchs, S. (2019). Financial recovery schemes in Austria: How planned relocation is used as an answer to future flood events. *Environmental Hazards*. doi:10.1080/17477891.2019.1665982

Thomalla, F., Lebel, L., Boyland, M., Marks, D., Kimkong, H., Tan, S. B., & Nugroho, A. (2018). Long-term recovery narratives following major disasters in Southeast Asia. *Regional Environmental Change, 18*(4), 1211–1222. doi:10.1007/s10113-017-1260-z

Tsvetkova, O., & Randhir, T. O. (2019). Spatial and temporal uncertainty in climatic impacts on watershed systems. *Science of the Total Environment, 687*, 618–633.

Xu, X. (2019). Prospects for disaster management in China and the role of insurance. *Environmental Hazards, 18*(5), 383–399. doi:10.1080/17477891.2019.1609404

Zavar, E. M., & Schumann III, R. L. (2020). Post-disaster communalism: Land use, ownership, and the shifting 'publicness' of urban space in recovery. *Environmental Hazards, 19*(4), 398–416. doi:10.1080/17477891.2019.1690969

OPEN ACCESS

Measuring social equity in flood recovery funding

Christopher T. Emrich, Eric Tate, Sarah E. Larson and Yao Zhou

ABSTRACT

Deconstructing causal linkages between place attributes and disaster outcomes at coarse scales like zip codes and counties is difficult because heterogeneous socio-economic characteristics operating at finer scales are masked. However, capturing detailed disaster outcomes about individuals and households for large areas can be equally complicated. This dichotomy highlights the need for a more nuanced and empirically-driven approach to understanding financial disaster recovery support. This study assessed how social characteristics influenced federal disaster recovery support following the 2015 South Carolina floods. Ordinary linear and spatial regression models provided a mechanism for pinpointing statistically significant links between individual/compound vulnerabilities and resource distribution from four federal disaster response and recovery programmes. The study makes two unique contributions. First, exploration of how social characteristics influence recovery support is a critical, yet understudied path toward fair and equitable disaster recovery. Second, finer scale inquiry across a large impact area is rare in quantitative case studies of US disasters. While we found flood recovery assistance to be strongly associated with physical damage overall the relationship was more tenuous in places with higher social vulnerability. Results indicate that future disaster recovery programs focusing on both physical damage and social vulnerable would lead to a more equitable disaster recoveries. Findings provide new understanding of equity at the intersection of social vulnerability, impacts, and disaster recovery and showcase both best-practices and areas for programme improvements for future disasters.

1. Introduction

Vaulted into the US national consciousness by the disparate population impacts of Hurricane Katrina (Gabe, Falk, & McCarty, 2005), there has been growing recognition that

This article has been republished with minor changes. These changes do not impact the academic content of the article.

understanding and managing flood disasters requires more than simply characterising hydrological processes and built environment damage. Social inequities are also now understood to be significant drivers of flood exposure, adverse impacts, and recovery trajectories (Burton & Cutter, 2008; Hale, Flint, Jackson-Smith, & Endter-Wada, 2018; Koks, Jongman, Husby, & Botzen, 2015). These drivers are conceptually and empirically investigated in social vulnerability research, which examines how systemic social, economic, institutional, and political processes lead some population groups to face pervasive barriers accessing post-disaster resources. Links between social vulnerability and flood recovery for various disaster types suggest that social characteristics might play an increasingly important role in flood risk management (Burton, 2010; Cutter, Schumann, & Emrich, 2014; Dunning & Durden, 2007). Integrating understanding of social vulnerability with knowledge of environmental and economic processes offers a pathway toward more sustainable futures in the aftermath of disaster (Smith & Wenger, 2007). Disaster management under this paradigm implies a mindset rooted in social equity that emphasises vulnerable populations in the distribution of recovery resources.

Social equity in disaster recovery means that all people have full and equal access in resource distribution and opportunities that enable them to meet their needs. Yet, social vulnerability scholarship is replete with examples of barriers that preclude the realisation of social equity in planning, response, and recovery (Thomas, Phillips, Lovekamp, & Fothergill, 2013). For flood hazards, equity has received growing focus in examining the social dimensions of resilience (Comes, Meesters, & Torjesen, 2019; Doorn, 2017; Doorn, Gardoni, & Murphy, 2019), vulnerability (Collins, Grineski, & Chakraborty, 2018; Rufat, Tate, Burton, & Maroof, 2015), risk management (O'Hare & White, 2018; Thaler, Fuchs, Priest, & Doorn, 2018; Thaler & Hartmann, 2016), and justice (Kaufmann, Priest, & Leroy, 2018; Siders, 2019; Walker & Burningham, 2011). Media and government reports often describe disaster assistance in terms of total amounts authorised or allocated. A complimentary, but often overlooked dimension is distributive equity in disaster programmes: the logic underlying the relative distribution of post-disaster assistance and the associated implications for meeting people's needs during flood recovery (Muñoz & Tate, 2016; Vinik, 2018). However, social equity is not a principal factor in the distribution of US federal recovery dollars.

Four national programmes dominate the outlay of disaster recovery funds to households in the US: Individual Assistance (IA), the National Flood Insurance Program (NFIP), Small Business Administration (SBA) disaster loans, and the Community Development Block Grant-Disaster Recovery (CDBG-DR). Each differs in eligibility criteria, assistance types, and assistance timing. NFIP pays flood claims based on pre-disaster insurance coverage. IA grants and SBA loan availability require a Presidential Disaster Declaration, while CDBG-DR allocations require a special congressional authorisation after a disaster with large unmet needs remaining after resource allocation from other recovery programmes. Collectively, the four federal programmes form an umbrella of financial resources available to households following major flood disasters. These programmes allocate billions of dollars annually toward disaster recovery, but how equitable is the distribution?

Flood recovery is increasingly recognised as an important part of flood risk management yet one that remains understudied (O'Neill, 2018; Schanze, 2006). Focusing on disaster programmes for housing recovery, this study investigates two research questions: (1) What is the social equity of US post-disaster assistance?; and (2) How does this equity vary by federal programme? To empirically evaluate these questions, we use a case study of the 2015 South Carolina floods to determine the relationship between social vulnerability indicators and disbursements from four disaster recovery programmes. The remainder of this paper proceeds as follows. The next section explains connections between social equity and social vulnerability, and describes the major US programmes that distribute disaster recovery funds. Sections 3 and 4 describe outcome measures from the South Carolina floods and detail the methods we applied to determine the statistical relationship between social vulnerability indicators and recovery resource outlays. We present the results of the analysis in Section 5, and discuss the major findings and conclusions in Section 6.

2. Background

2.1 Social equity and social vulnerability

Environmental justice scholarship provides useful guidance for conceptualising social equity. The field has long organised research around forms of justice, which include distributive, procedural, social, corrective, and retributive types (Kuehn, 2000). Distributive justice is the right to equal treatment across demographic groups, and associated analyses typically quantify how environmental hazards vary across populations. With its emphasis on the notion that no population group should disproportionately benefit or be burdened by the distribution of environmental harms or resources, social equity strongly aligns with the distributive justice (Ikeme, 2003). Based on social equity principles, the distribution of disaster recovery resources should be based on fulfilling the needs of the affected. But who has the greatest needs after disaster?

For each of the major programmes for disaster recovery, households with the greatest need are largely identified based on physical damage. The identification generally occurs through field assessment of post-disaster damage and estimation of associated repair costs. Under this paradigm, households with damage to highly valued assets are prioritized in the distribution of recovery resources. But this design is unlikely to satisfy the post-disaster needs of socially vulnerable populations. These groups tend to experience greater proportionate losses (Fothergill & Peek, 2004), have less bridging and linking social capital, and face greater delays in receiving recovery dollars (Muñoz & Tate, 2016). To achieve social equity, criteria for defining post-disaster needs should expand beyond physical vulnerability to also include social vulnerability.

Social vulnerability is the heightened propensity of marginalised populations to suffer adverse disaster impacts. Such marginalisation is rooted in socially-stratified economic, social, and political power structures, which manifest as geographic and demographic disparities in wealth, education, and legal rights. Vulnerability researchers have long linked these disparities with differential capacity to prepare for, cope with, and recover from disasters (Wisner, Blaikie, Cannon, & Davis, 2004), with results dispelling the notion that disaster recovery occurs uniformly across all members of affected communities (Smith & Wenger, 2007; Tierney & Oliver-Smith, 2012). To quantitatively model processes of disparity, social vulnerability researchers have used empirical understanding about what makes people and places vulnerable, extracting associated socio-demographic variables and assembling them into indicators of social vulnerability. If tailored to flood recovery, social vulnerability indicators hold promise as measures of social equity.

Social vulnerability manifests differently across the landscape and may compound from underlying demographic characteristics (Emrich & Cutter, 2011; Morrow, 1999). Where one person or place might be vulnerable because of age or racial discrimination, a neighbouring person or place might be vulnerable (or not) due to different characteristics or circumstances (e.g. poverty, renter, residing in a mobile home). Previous studies have found that intersections of vulnerability characteristics are linked to adverse outcomes beyond those associated with individual constituent drivers. For example, Elliott and Pais (2006) in their study of Hurricane Katrina found that low-income Black populations experienced greater post-disaster job loss than low income populations or Black populations alone. Yet despite conceptual and empirical understanding of social vulnerability as compounding, most quantitative studies treat social vulnerability drivers as discrete (Ryder, 2017). Few have explored the interactions between compounded social vulnerabilities and disaster outcomes (Rufat, Tate, Emrich, & Antolini, 2019).

2.2 Disaster recovery programs for housing

FEMA's National Disaster Recovery Framework clarifies the roles and responsibilities for stakeholders in recovery, recognising first that recovery is a continuum (FEMA, 2016a). Because disasters impact some segments of the population more than others, the ability of an individual or community to accelerate the recovery process is highly dependent on resources (insurance, savings, federal recovery support, non-profit assistance) in every disaster phase. Delivery of the right post-disaster resources to populations in need can significantly reduce recovery time and cost. The distribution of these resources can occur over short time frames after the disaster (days to months) through IA housing grants, NFIP insurance payments, and SBA home loans, as well as in delayed fashion (months to years) through CDBG-DR rebuilding grants (Figure 1).

IA is a grant programme intended to address immediate needs of uninsured and underinsured disaster survivors. IA accounted for 14% of federal disaster spending from fiscal

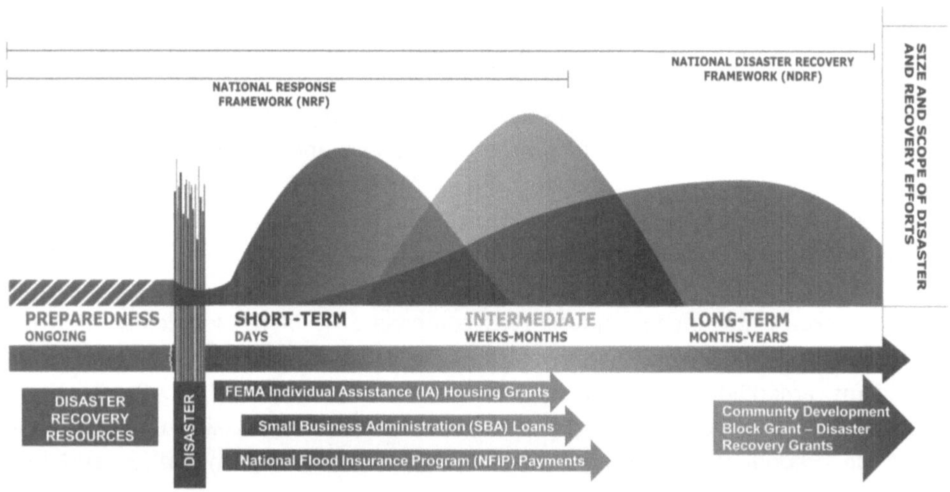

Figure 1. Recovery continuum and federal recovery programmes (adapted from FEMA 2016b).

years 2007–2016 under the Stafford Act (Reese, 2018), and 4.7 million people applied to the programme in 2017 (GAO, 2018). Among the six IA subprograms, the Individuals & Housing Program is the largest and provides funds for housing repair, short-term lodging reimbursement, personal property replacement, medical care, moving & storage, and transportation costs (FEMA, 2016b). Funding maximums are indexed to inflation, and in 2018 were $33,000 for households and $8,500 for individuals (Reese, 2018), although average payouts are typically much lower. Disaster survivors have up to 60 days after a disaster declaration to apply (Figure 1), and to be eligible must be US citizens or qualified residents, have expenses and needs caused the disaster, and document that other sources of disaster assistance are insufficient to meet needs (Lindsay & Reese, 2018). IA eligibility is also subject to income screening, as families with incomes above 1.25 times the federal poverty level (1.5 times for individuals) are instead referred to the SBA disaster loan programme (Lindsay & Reese, 2018). Recent studies examining the relationship between IA and demographic characteristics found social vulnerability indicators to be significant predictors of the number of applicants, physical damage, and aid granted (Grube, Fike, & Storr, 2018; Rufat et al., 2019). Based on IA's focus on uninsured and poorer households, we expect moderate overlap between IA grantees and socially vulnerable populations.

The SBA disaster loan programme provides resources to uninsured and underinsured households and businesses. Although managed by the Small Business Administration, 80% of the loan funds are allocated to households (Kreiser, Mullins, & Nagel, 2018). Homeowners can borrow up to $200,000 to repair or restore primary residences to pre-disaster condition, while both renters and homeowners can borrow up to $40,000 to replace damaged personal property (e.g. cars, clothing, furniture). Eligibility requires verification of disaster-related losses, satisfactory credit, and ability to repay the loan (Lindsay & Reese, 2018). The availability of SBA loans for households is triggered by a disaster declaration that includes IA. Because SBA loan recipients incurred underinsured losses yet also have sufficient resources to be deemed credit worthy, we expect a weak positive relationship with socially vulnerable populations.

NFIP provides subsidised flood insurance to property owners in participating communities that map high risk areas and regulate floodplain development. The NFIP coverage ceiling is $250,000 for residential properties, and approximately 90% of residential policies cover single family dwellings (FEMA, 2018). Thus, NFIP is a potentially large source of disaster resources to flood-impacted households. However, property owners in non-participating communities are ineligible to purchase flood insurance. A recent comparison of household income and NFIP found that those without NFIP coverage had a median income of $40,000 (University of Maryland, Center for Disaster Resilience, and Texas A&M University, Galveston Campus, Center for Texas Beaches and Shores, 2018) compared to the national median income of $61,372 (Fontenot, Semega, & Kollar, 2018). This is important because low-income households comprise 51% of households in high risk areas (FEMA, 2018). Meanwhile, a study of Hurricane Harvey found that Black and Hispanic flood victims carried flood insurance at lower rates than whites (Hamel, Wu, Brodie, Sim, & Marks, 2017). We therefore expect NFIP payouts to be lower for socially vulnerable areas due to the predominance among the underinsured of renters, minorities, and low-income residents.

CDBG-DR provides funds for short-term and long-term disaster recovery. Because disbursement of CDBG-DR funds requires a special congressional authorisation, it is often several months post-disaster before programme funds reach affected communities. As such, the programme helps fill gaps in resident needs that remain unmet after NFIP, IA, SBA assistance (SCDRO, 2017). CDBG rules allow for a wider array of permissible expenditures than IA and SBA, so communities have flexibility to tailor spending based on local needs. Receiving states and communities must allocate at least 70% of funds to activities that principally benefit low-to-moderate income households, although waivers have at times relaxed this requirement to 50% for disaster recovery (Boyd & Gonzales, 2011). Low-to-moderate income is defined by CDBG as households below 80% of the median income for their area. Given the explicit programme emphasis on low-income households, we expect CDBG-DR to have a moderate positive relationship with socially vulnerable populations.

Social vulnerability indicators are routinely mapped to describe geographic patterns of social vulnerability and are increasingly applied statistically to analyse disaster outcomes. A far less employed application of social vulnerability indicators is as tools for evaluating disaster programmes. Given that socially vulnerable populations are disproportionately impacted by disasters, does resource distribution through the leading federal disaster programmes accrue to socially vulnerable groups? As such, this article focuses on the following two research questions: (1) What is the social equity of US post-disaster assistance?; and (2) How does this equity vary by federal programme?

3. The 2015 South Carolina flooding and severe storms disaster

In October 2015, South Carolina experienced unprecedented flooding as a result of heavy rainfall across the State resulting from Hurricane Joaquin (NWS 2016). Abundant rainfall in September left saturated soils with little absorptive capacity. Runoff from Joaquin caused rivers to rise quickly and fill water behind a large number of the state's 2500 dams. An estimated 47 dams statewide breached, flooding downstream residents and businesses (Sasanakul et al., 2017), while more than 500 roads were closed, with some collapsing under the weight of saturated soils. The flooding caused loss of life and extensive damage to dams, bridges, roads, homes, and businesses. The rapid onset resulted in many residents requiring swiftwater rescue, and the damage impacted utilities, wastewater treatment systems, and drinking water treatment and collection systems. The severe storms and flooding prompted a Major Presidential Disaster Declaration opening the flow of federal disaster recovery resources (IA and SBA) to 25 counties across the state (Figure 2).

More than 101,000 applicants filed for IA for the South Carolina's 2015 flooding disaster (Figure 3(A)). More than 76% are homeowners and about 24% were renters residing mainly in single family homes, duplex units, and mobile homes (SCDRO, 2017). According to the Vulnerability Mapping and Analysis Platform (VMAP, 2019), only about 28% of the IA applicants received disaster recovery resources, a denial rate among the top ten in recent similar disasters. Less than half as many flood victims applied for SBA disaster loans during the weeks following the floods (Figure 3(B)), but only 9.6% of SBA applicants ultimately received loans. NFIP provided payouts to 84% of the 5504 flood victims who filed flood insurance claims (84%) (NFIP, 2016). A CDBG-DR programme for the disaster was

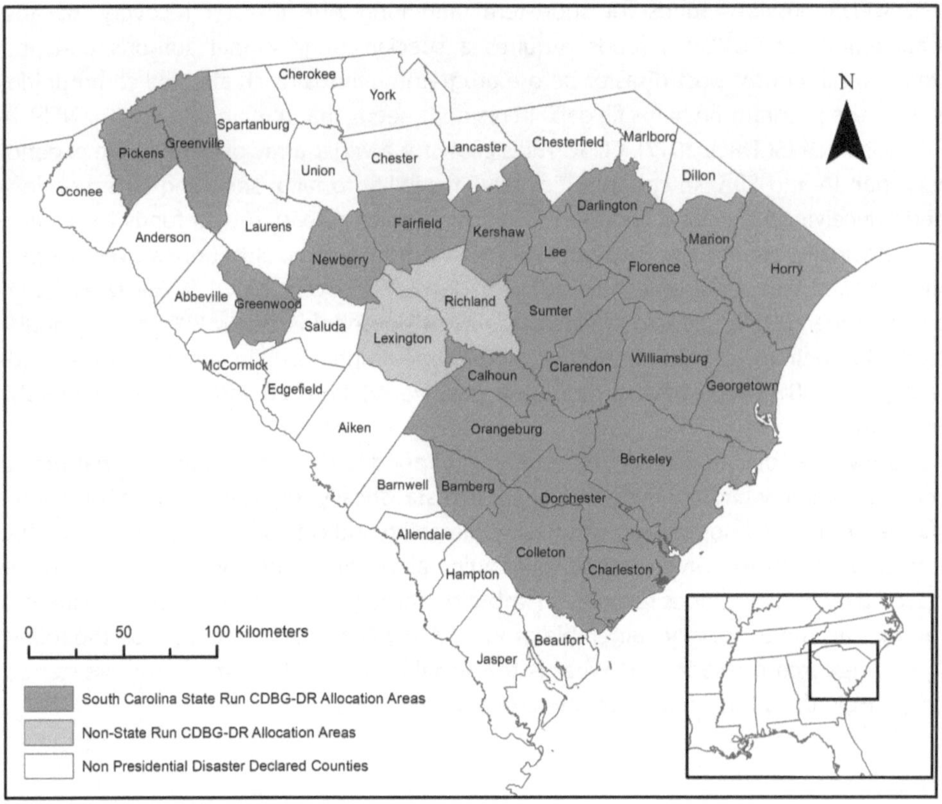

Figure 2. South Carolina study area.

implemented in Spring 2016, and has provided resources to approximately 25% of more than 8000 applicants (25%) to date (Figure 3(D)) (SCDRO, 2019).

4. Methods and data

We compiled recovery programme information from the 2015 South Carolina floods, and demographic variables from the US census. The objective was to statistically explore linkages among univariate and compound social vulnerability, disaster losses, and recovery resources. To do so, we constructed multivariate regression models that use program-specific funding totals as dependent variables and social vulnerability indicators as predictors while controlling for loss.

4.1 Flood outcome measures

Outcome variables for the analyses include loss indicators and programme support indicators for real property (building) and personal property (contents) (Table 1). We calculated loss and support indicators separately by recovery programme because each programme accounts for losses differently.

We computed the dependent and control variables for each programme as sums of multiple damage/loss and support measures (Figure 4). Loss and payouts (grants, insurance, or

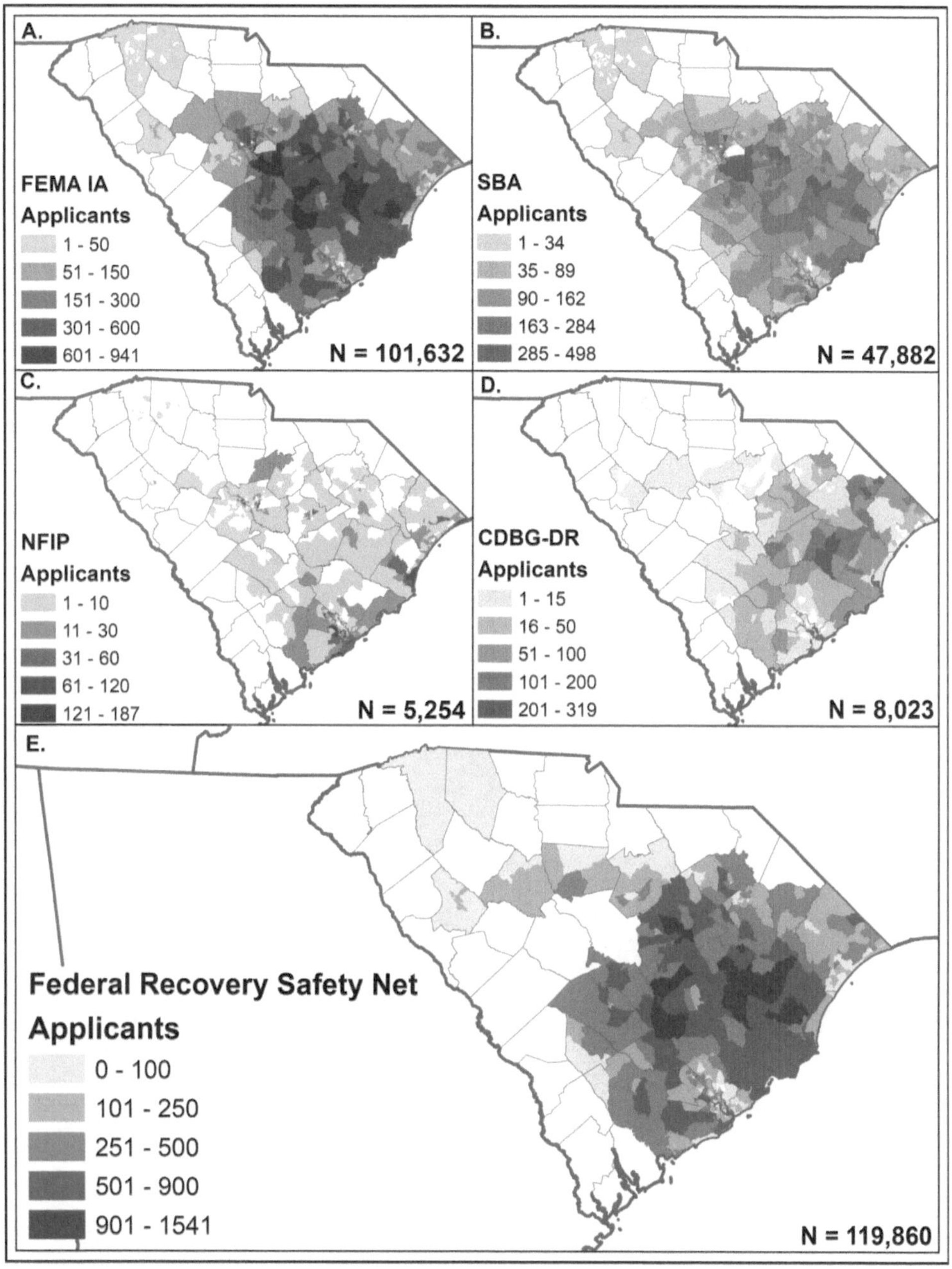

Figure 3. Disaster survivors by recovery support programme – IA (A); SBA (B); NFIP (C); CDBG-DR (D); Combined programmes (E).

loans) to property owners dominate the totals in comparison to losses and payouts for personal property (contents). As such, property losses and payouts appear to be the most useful variable for understanding the driving forces behind disaster equity across the study area. We calculated the program-specific measures of disaster loss as follows:

Table 1. Disaster outcome variables.

Programme	Type	Variable	Description (US dollars)	Source
IA	Dependent†	Housing Assistance (HA)	HA building repair/ replacement grant	FEMA (2016c)
	Dependent†	Other Needs Assistance (ONA)	ONA contents replacement grant	
	Control‡	FEMA Full Verified Loss for Real Property	Building damage	
	Control‡	FEMA Full Verified Loss for Personal Property	Contents damage	
SBA	Dependent†	Current Real Property Loan Amount	Current loan amount for building recovery (manufactured, repair/ replace, reconstruction, relocation)	SBA (2019)
	Dependent†	Current Contents Loan Amount	Current loan amount for contents replacement	
	Control‡	SBA Verified Real Property Loss	Sum of building damage (manufactured, repair/ replace, reconstruction, relocation)	
	Control‡	SBA Verified Contents Loss	Contents damage	
NFIP	Dependent†	Building Payments	Insurance payouts for building repair, replacement	NFIP (2016)
	Dependent†	Contents Payments	Insurance payouts for contents replacement	
	Control‡	Building Loss	Building damage	
	Control‡	Contents Loss	Contents damage	
CDBG-DR	Dependent†	Final Amount	CDBG-DR funds spent on repair/replacement	SCDRO (2019)
	Control‡	Estimated Rebuilding Amount	Estimated building damage	

‡Measure of loss.
†Measure of support.

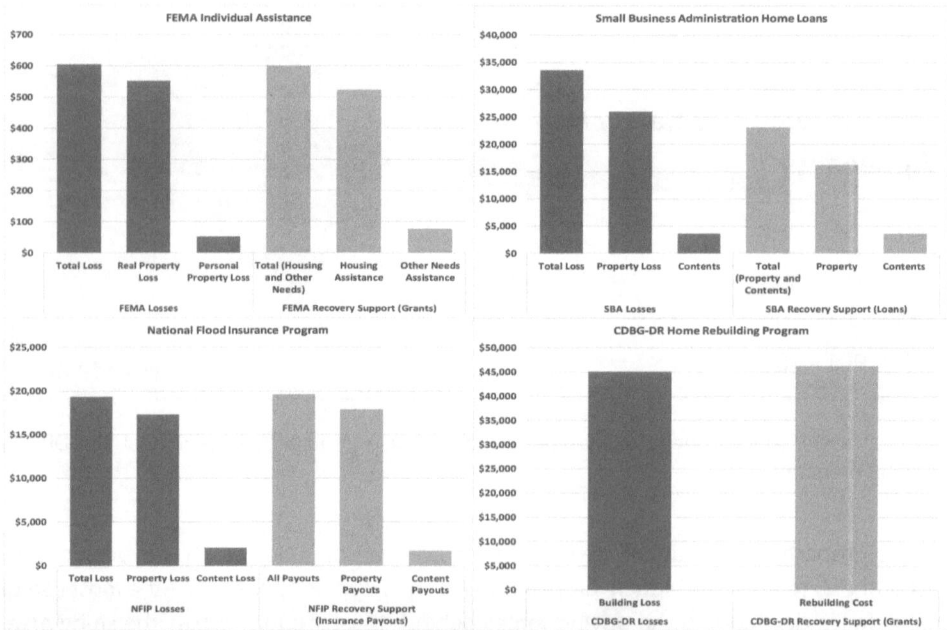

Figure 4. Loss and support variables by federal disaster recovery programme.

Table 2. Social vulnerability indicators.

Driver	Indicator	Rationale for Inclusion	Sources
Housing tenure	% Renter Households	Less access to post-disaster housing programmes; lower post-flood return rate; difficulty exiting rental agreements in damaged homes	Finch, Emrich, and Cutter (2010), Kamel (2012), NASEM (2019)
Financial capital	Per Capita Income	Limited recovery resources and options for post-flood housing; higher post-flood health impacts; disproportionately reside in flood-prone areas; less likely to carry flood insurance	Green, Bates, and Smyth (2007), Masozera, Bailey, and Kerchner (2007), NASEM (2019)
Race	% Black Population	Higher death and injury rates; negative post-flood health outcomes; less likely to carry flood insurance; lower trust in authority for post-flood assistance; higher employment loss	Elliott and Pais (2006), Li, Airriess, Chen, Leong, and Keith (2010), Hamel et al. (2017)
Language proficiency	% Speak English as a Second Language: Not Well or Not At All	Difficulty accessing recovery assistance from governmental and nongovernmental sources; fewer adverse post-flood health outcomes	Collins, Jimenez, and Grineski (2013), NASEM (2019)
Housing quality	% Mobile Homes	Added repair difficulties; focus of the SC CDBG-DR implementation.	SCDRO (2017)
Age	Age Under 5 and Over 65	Negative post-flood health outcomes; lower recovery rates	Collins et al. (2013), Muñoz and Tate (2016), NASEM (2019)
Employment	Service Sector Employment	Sector specific employment (service industry) is particularly vulnerable to disasters. Lack of disposable income following disasters decreases the need for often low-paid service sector jobs.	Hewitt (1997), Puente (1999), Heinz Center (2000)

- IA loss = sum of FEMA verified loss for personal and real property
- SBA loss = sum of SBA verified loss to property (repair and replacement), contents damage, relocation expenses, damage to other structures, manufactured home damage and debris removal
- NFIP loss = sum of the building and contents damage as reported by NFIP claims adjusters
- CDBG-DR loss = final rebuilding cost or estimated rebuild cost (when the final amount was not available)

4.2 Social equity indicators

We selected seven social vulnerability variables (Table 2) at the census tract level using data from the 2011–2015 American Community Survey of the US Census Bureau. The census tract scale is the smallest geographic unit for which demographic variables are considered to be statistically reliable (Spielman, Folch, & Nagle, 2014). The variable selection was guided by findings from previous empirical studies that examined links between social vulnerability characteristics and flood recovery. We used a smaller than typical (VMAP, 2018) subset of vulnerability indicators for two purposes: First, the variables align with the stated goals, event type, and distribution methods of each recovery programme. Second, the small indicator set maintains statistical power in the analysis, enabling exploration of interaction effects as a proxy for compound vulnerabilities.

Table 3. Pearson correlations of social vulnerability indicators.

	% Renters	Per Capita Income	%Black	%ESL	%Mobile Homes	%Population under 5 or over 65	%Service Industry Employment
% Renters	1.00						
Per Capita Income	−0.2767***	1.00					
% Black	0.3456***	−0.5257***	1.00				
% ESL	0.0968***	−0.2163***	0.0430	1.00			
% Mobile Homes	−0.3391***	−0.3678***	0.0935***	0.1437***	1.00		
% Population under 5 or over 65	−0.3433***	0.2122***	−0.0719**	0.0013	0.0459	1.00	
% Service Industry Employment	0.3795***	−0.4851***	0.4102***	0.4105***	0.0033	−0.0569	1.00

***$p < 0.01$, **$p < 0.05$.

The Pearson correlation coefficients of the social vulnerability indicators range in absolute value from 0.0013 to 0.5257, suggesting they represent distinct dimensions of social vulnerability (Table 3). We examined compounding vulnerability by analysing first-order interactions among programme-specific loss indicators (control variables) and three of the social vulnerability predictors: renters, wealth, and race. We selected only these predictors due to their empirical importance and to retain statistical power across models. The remaining four social vulnerability variables were included in the regression models as individual predictors, but not interacted with the loss indicator.

4.3 Multivariate regression models

We constructed five multivariate regression models to assess social equity in the allocation of flood recovery resources. Each model used programme support as the outcome variable, social vulnerability indicators as predictors, and a loss measure as control. The regression models are a combination of ordinary least squares (OLS) and spatial autoregressive (SAR) models. To determine if a given OLS model contains residual spatial autocorrelation, we computed the Moran's I statistic on the OLS residuals. When significant, we applied the Lagrange Multiplier test to determine the dominant form of spatial effects (heterogeneity or dependence), and then applied the appropriate spatial regression model (Spatial Error or SAR). For both the Moran's I statistic and the spatial regression modelling, we generated the weights matrix based on a K-nearest neighbour schema ($K = 10$).

The outcome indicators were collected by the federal agencies at the household scale, but we aggregated them to the census tract scale to match the scale of the census demographic variables. The aggregation means that the statistical linkages between social vulnerability variables and outcomes apply to the dominant demographic characteristics of the containing census tracts rather than those of the recipient households. Were the federal programmes to collect detailed demographic data during future distribution of recovery resources, associated equity analysis could produce finer-grained results.

Table 4 provides descriptive statistics for the outcome and predictor variables, including the number of census tracts for which each variable was available. Variations in

Table 4. Descriptive statistics for disaster recovery programmes.

Statistic	*N*	Mean	St. Dev.	Min	Median	Max
Disaster Outcomes						
IA Support	764	$600	$650	$0	$454	$5584
IA Loss	764	$634	$810	$0	$454	$10,687
SBA Support	586	$19,256	$15,895	$0	$17,628	$190,900
SBA Loss	586	$30,422	$22,401	$0	$25,127	$215,871
NFIP Support	437	$17,908	$32,319	$0	$10,188	$500,000
NFIP Loss	437	$19,342	$38,495	$0	$10,718	$654,401
CDBG Support	230	$46,167	$18,588	$3185	$45,680	$121,288
CDBG Loss	230	$45,097	$17,969	$3,185	$44,035	$121,288
Total Support	582	$52,079	$45,508	$0.00	$39,673	$576,144
Social Vulnerability						
% Renters	764	29.15%	16.95%	0%	25.42%	100%
Per Capita Income	764	$24,491	$11,424	$0.00	$21,865	$127,428
% Black	764	30.65%	24.57%	0%	23.81%	97.69%
% ESL	764	20.38%	21.15%	0%	14.69%	100%
% Mobile Homes	764	16.18%	17.57%	0%	8.64%	81.75%
% Population under 5 or over 65	764	21.22%	6.29%	0%	20.73%	57.80%
% Service Industry Employment	764	19.29%	7.66%	0%	18.00%	55.56%

programme-specific enrolment (e.g. some programmes did not have applicants in some census tracts) resulted in a different set of census tracts for each regression model. Among the programmes, CDBG-DR funds were distributed in the lowest number of tracts, but had the highest average allocation. By contrast, IA resources were distributed to the greatest number of tracts, but had by far the lowest average allocation. Lower average and median percentages across the study area for most of the social vulnerability variables stands in stark contrast to a few instances (% Black, % ESL, and % mobile home) where the maximum percentage in at least one census tract is nearing 100%.

Model 1 is an SAR model with IA support as the dependent variable and equity variables as predictors. Positive spatial autocorrelation between outcome and predictor variables warranted use of a SAR lag model with maximum likelihood. The SAR model tested relationships between average IA grants and the seven social vulnerability predictor variables, both independently and with interactions among the three selected social vulnerability indicators, while controlling for IA losses.

Model 2 analyses the distribution of recovery dollars through the SBA loan programme. We used OLS regression after finding no spatial collinearity between the dependent and independent variables. Average SBA loan amounts were regressed against the seven independent measures of social vulnerability, three of which were also interacted to identify compound vulnerability influences on SBA support controlling for SBA's measure of loss.

Models 3 and 4 examine NFIP and CDBG-DR data, respectively. SAR modelling was not possible for either programme due to substantial spatial disconnectivity between tracts where funds were allocated (Figure 3). Consequently, we used OLS regression in the analysis for these two programmes. For each, the average support was tested against the seven independent measures of social vulnerability, three of which were selected for interaction among themselves and with the remaining four independent variables, and one control variable of average programme specific damage.

Model 5 is a combined analysis to explore the aggregate 'federal disaster recovery safety net,' in which we calculated a measure of total household support as the sum of average support from each of the four programmes. Unfortunately, linking specific assistance recipients across federal programmes is not possible because common identifiers are not collected by each recovery programme. As such, the total safety net model reflects aggregate resources disbursed to places rather than specific households in places. Were common identifiers available, it would allow for an improved measure of household recovery outcomes. Model 5 analysed the same set of predictor variables as in Models 1–4. We identified spatial autocorrelation in the model data and therefore selected SAR over OLS.

Selecting an appropriate loss measure as the control variable emerged as an important methodological decision, because combining losses from each programme would likely result in double counting. A greater understanding of each programme's loss measure is imperative for making an informed decision in this regard. NFIP only measures losses for individuals with insurance who filed a claim ($N = 5254$), which excludes uninsured populations. CDBG-DR measures loss as a function of rebuilding costs for those in the programme ($N = 8023$), suggesting the possibility of missing damage costs. IA only evaluates losses to essential living spaces, excluding extra bedrooms or living areas inside a home from damage estimates (FEMA, 2019). So although IA includes the highest number of applicants and census tracts across the study area, it underestimates total loss. Because SBA measures all losses to real property and personal property and includes more applicants and census tracts than NFIP or CDBG-DR, we judged the SBA loss measure to be the best suited for the combined analysis. As a result, Model 5 only includes census tracts where SBA was allocated.

5. Results

Pearson correlations between each individual programme's measure of recovery support and loss (Table 5) show high coefficients for NFIP and CDBG-DR. This strong positive relationship is expected because both programmes focus on repairing or replacing based on total loss. Furthermore, because NFIP is an insurance programme, the average insurance support at the tract level received should be extremely similar to the measure of loss of the claims adjuster. Like NFIP, the nature of CDBG-DR as a single-family repair/replace programme suggests that losses and support will be extremely similar. There are also strong associations between loss and support within the IA programme. The lower correlation value for the SBA programme may be related to ineligibility of applicants or unwillingness to take on a recovery loan.

Table 6 presents the results of the multivariate regression for all five models. Non-significant findings, represented by the empty cells (Table 6) as well as all the variables (Table 2) that had no influence in any model and are not included in results. For each of the programme-specific models, the loss measure is statistically significant in predicting the various forms of support. This suggests that the loss variable is appropriately controlling

Table 5 . Correlations between dependent (loss) and independent (support) variables.

		Loss			
		CDBG	NFIP	SBA	IA
Support	CDBG	0.9842***			
	NFIP		0.9867***		
	SBA			0.5585***	
	IA				0.8492***

***$p < 0.01$, **$p < 0.05$.

Table 6. Significant regression results for each recovery programme and the composite Federal disaster safety net†.

	Influence on Prgram Specific Federal Disaster Recovery Support				
	FEMA IA Grants	SBA Loans	NFIP Payouts	CDBG-DR Grants	Federal Disaster Safety Net
Univariate Drivers of Disaster Losses and Funding					
Programme Specific Loss† Per Capita	1.04*** (0.08)	−0.29*** (0.11)	0.96*** (0.05)	1.07*** (0.07)	
Per Capita Income	0.02*** (0.00)	0.74** (0.34)			
% Renters			−20,486.9** (9975.40)		
% Black		−61,082.98*** (20,907.17)			
% Service Sector Employment				82,196.59*** (24,141.02)	
Compounded (Multi-Variate Drivers of Disaster Losses and Funding)					
Programme Specific Loss† and Per Capita Income	−0.00*** (0.00)	0.00*** (0.00)	−0.00** (0.00)		0.00** (0.00)
Programme Specific Loss† and % Renters			−0.38*** (0.04)	0.22** (0.09)	0.32** (0.12)
Programme Specific Loss† and % Black		1.04*** (0.15)	0.23** (0.04)		
Per Capita Income and % Under 5 or over 65	−0.03** (0.01)				
% Renters and % Under 5 or over 65	2054.70** (1024.9)				
% Speaking English Not Well or Not at All and % Renters			15,307.17** (7541.10)		
% Speaking English Not Well or Not at All and % Black			−10,385.07** (5051.74)		
% Speaking English Not Well or Not at All and % Black				−51,146.43*** (19,237.80)	
Per Capita Income and % Renters				−1.02** (0.51)	
% Black and % Mobile Homes					26,318.00*** (12,712.00)
Constant	−566.95** (268.15)	51,057.08*** (12,988.78)	4324.52 (4634.83)	−18,780.85 (11,339.62)	7833.40 (9554.00)
Observations	764	586	437	230	582
Log Likelihood	−5500.93	−6340.23	−4226.94	−2161.00	−6134.07
Adjusted $R2$ /(ρ)	(0.147***)	0.393	0.985	0.972	(0.439***)

***$p < 0.01$, **$p < 0.05$, †SBA losses used for safety net loss models.

variations in loss that would explain the variation in average support received. The OLS and SAR models were structured such that positive beta coefficients in regression results indicated higher amounts of recovery support and larger beta coefficients indicate larger changes in amount received in relation to each single unit change in the predictor variable. Specific model results are discussed below.

5.1 Equity in IA grants

Univariate interactions between predictor variables and IA outcome measures indicated that average loss per applicant per census tract statistically and significantly explained IA support. Here, as losses increased, support also increased. Given this is a control variable, we expected this relationship to be statistically significant. Among the three social vulnerability indicators of greatest interest (% renter, % Black, and per capita income) only income was independently statistically significant in explaining average IA support received per census tract. As income increased, average IA support increased, signifying that census tracts with higher per capita income received higher IA funding. None of the four other indicators of social vulnerability were independently statistically significant (Table 6).

We analysed compound vulnerabilities through variable interactions among the social vulnerability indicators. Two joint interactions are statistically significant for IA: (1) the interaction between age and per capita income and (2) the interaction between age and renters. These findings suggest that in census tracts with higher than average percentage population under 5 and over 65 and higher than average income levels, a lower amount of IA support was provided than in census tracts with below average levels of percentage population under 5 and over 65 and below average income. This suggests that age is an important indicator of social vulnerability and could adversely impact IA support even in cases with higher income levels. For the interaction of age and renters, census tracts with higher than average populations under 5 or over 65 and higher than average renters, a larger amount of IA support was provided than to those census tracts with below average levels of both variables. This suggests IA is reaching socially vulnerable populations in this specific joint interaction.

5.2 Equity in SBA disaster loans

Assessing univariate drivers of SBA support yielded several significant results. Similar to IA, the loss variable is statistically significant (Table 6). However, the directionality of the sign is opposite what we would expect, suggesting that places with higher overall losses received fewer SBA loans. This negative relationship might be because SBA provides loans that must be repaid rather than grants. In these places with higher losses, disaster victims might be either exhibiting higher rates of ineligibility or lower willingness to accept loans as recovery support. Among the social vulnerability indicators, income was negatively correlated with average SBA support, suggesting that as average income increased, the amount of SBA average support decreased. Conversely, as the Black population increased, average SBA support decreased, signifying that census tracts with higher percentage of Black populations were adversely aided by this programme. No other univariate indicators of social vulnerability were statistically significant.

Interactions among independent variables as predictors of SBA support produced several results. Average loss per applicant per census block was statistically significant when interacted with two of the three of the social vulnerability indicators of greatest interest (per capita income and % Black). In places where losses were higher than average and the Black population or income was higher than average, SBA support was also higher than average. These findings suggest that SBA is reaching non-white populations in census tracts with higher losses and is also serving places where losses and income are generally higher. Among the other measures of social vulnerability, no joint interactions are statistically significant.

5.3 Equity in NFIP insurance payouts

Similar to the findings with the IA programme, the control variable was statistically significant. It is thus acting as an appropriate control for the variation in loss that would explain the variation in average support received at the census tract level (Table 6). Among the univariate social drivers of NFIP support, only renters produced a statistically significant relationship. The high negative coefficient suggests that places with a higher percentage of renters received less NFIP support. This finding makes logical sense, as NFIP is designed for homeowners. No other social vulnerability indicators were individually statistically significant.

Three compounded (multivariate) relationships were revealed through the OLS related to losses and social vulnerability. Losses interacted with income producing a negative influence on NFIP support. This result suggests that where income and loss were higher than average, NFIP support was lower. Similarly, in places where average loss and renters are higher than average, NFIP support was lower. Conversely, in places where the Black population and loss were higher than average, NFIP support was higher.

Two joint interactions of only social vulnerability predictors were statistically significant in understanding NFIP support: (1) the interaction of language proficiency and percentage Black and (2) the interaction between language proficiency and renters. In places with higher than average language proficiency and higher than average Black population, NFIP support was lower than tracts with below average levels of each group. This finding suggests that NFIP is not reaching census tracts with this joint measure of social vulnerability. Conversely, places with both higher than average language proficiency and higher renters received higher NFIP support. This finding is interesting as NFIP building reconstruction loans are only available to homeowners, indicating that these places may also have high numbers of owner-occupied housing.

5.4 Equity in CDBG-DR rebuilding grants

The statistical significance of CDBG-DR loss in predicting support indicates the loss variable is an appropriate control for the variation in loss that would explain the variation in average support received. Interestingly, only one of the social vulnerability indicators (% service sector employment) was significantly explained average CDBG-DR support. As the percentage of individuals employed within the service sector increased, the average support also increased.

Compounding vulnerabilities produced three significant results in relation to CDBG-DR support. First, places with higher than average programme specific losses and higher than average renters received higher levels of CDBG-DR support. Second, census tracts with higher than average Black populations and higher than average language proficiency received less CDBG-DR support. Finally, tracts with both higher than average renters and income received less CDBG-DR support.

5.5 Equity in composite program support

The composite model combined support from all four programmes, and represents our attempt to model the federal safety net for flood recovery. Unlike the programme-specific models, no beta coefficients were statistically significant for any of the independent or control variables. However, several multivariate interactions produced significant influence on total disaster recovery support. Among these, average applicant loss was statistically significant in interaction with income and renters. These findings suggest that in cases where income and loss were higher than normal, total support was higher. In addition, in census tracts where renters and loss were higher, total average support was higher. The interaction term for mobile homes and percentage Black was also statistically significant. Census tracts with higher than average mobile homes and higher than average percentage Black received higher overall support, signifying the combined social safety net is reaching this joint interaction of social vulnerability.

6. Discussion and conclusions

Vulnerability manifests itself differently across landscapes. The intersection (or compounding) of univariate vulnerability measures (e.g. poverty, race, employment) creates unique situations for every place. These differences should be considered when analysing disaster outcomes. Specifically, considering compounding vulnerabilities in disparity and equity analysis aligns more with conceptual understanding of social vulnerability processes than using only univariate indications of vulnerable populations or indices. Analysis of the compound effects of vulnerability on impacts and outcomes enables understanding of the multivariate nature of social vulnerability drivers. This analytic approach produced findings for each recovery programme and the aggregate federal disaster safety net. Included in these noteworthy findings are non-significant results, significant equitable results, significant inequitable results, and significant results that are neither equitable nor inequitable. The model results did not always support our hypotheses, but did point toward both success and challenges for equitable disaster recovery planning and implementation.

Disaster outcomes are often difficult to predict in the ways that vulnerability literature conceptualised and theorises.[1] Disaster losses, in this case, are not always easily explained by univariate social measures or compound vulnerability indicators. The key variables drawn from the literature did not always perform in expected ways when assessing disaster specific outcomes. No programme is 100% equitable or 100% inequitable. Our analysis identified a balance between equitable outcomes and opportunities for improvement in serving the most vulnerable populations. Although the "balance" between equitable and inequitable outcomes is clear across all programs, one should understand that these do not cancel each other out. Even one instance of inequitable recovery fund distribution should be considered problematic and corrected by respective programs or funding entities in future disaster recoveries.

The IA programme is balanced in that both equitable and inequitable distribution of funds was related to impacts and outcomes. As expected, places with higher damage levels received more IA funding. When accounting for compound vulnerabilities, we find that places with higher losses and higher income levels received less IA recovery support, a finding in line with a general consensus that these populations can likely recovery without additional government assistance. Furthermore, and perhaps contrary to expectation, compound vulnerability analysis shows that places with higher average losses and higher % renters received higher amounts of IA recovery support. IA support was not systematically linked with higher or lower social vulnerability in the South Carolina case. In this regard, our hypothesis of a strong overlap between IA funding and socially vulnerable populations was not fully supported.

The SBA loan programme is largely equitable across the impact area with at least three indications of positive equity and only one indication of negative equity. The negative coefficient on SBA fund receipt in relation to SBA average loss indicates that factors associated with SBA loan receipt are more influential than SBA losses alone. Unfortunately, this inequitable outcome points to a potentially troubling finding: places with higher percentages of Black populations received significantly less SBA home loan support than places with lower percentages. Overall, the links between SBA support and social vulnerability align with our hypothesis of a weak relationship. We expected a positive association between vulnerable populations and SBA, but found a mixture of positive and negative outcomes.

NFIP was also balanced in its distribution of funds between equitable and inequitable. Here, places with higher percentage renter populations were less likely to receive funds. This finding expected because NFIP is designed for single family homeowners. When assessing compound vulnerabilities, we find that places with both higher than average NFIP losses and higher renter populations received lower amounts of NFIP payouts. Our hypothesis of a negative relationship between NFIP and social vulnerability is largely supported by our models with only one significant exception. However, we find that a combination of race and low English language proficiency result in lower NFIP payouts.

CDBG-DR provided more recovery support to places with higher per applicant losses than places with lower losses. The state-run programme also provided some support in ways that were not specifically outlined in their programme developments. Supporting our hypothesis of moderate associations between CDBG-DR support and socially vulnerable populations, the single-family home rebuilding programme provided significantly higher levels of funding to places with higher percentages of service sector employees over places with lower percentages. Furthermore, places with higher losses and higher percentages of renters received more funding. These positive unintended consequences of the programme support a broader conceptualisation of social vulnerability beyond individual (rich/poor) indicators. However, places with higher percentages of service sector employees and

Black populations received significantly less CDBG-DR support than places with both lower service industry employment and lower percentages of Black people.

This paper contributes significantly to disaster recovery literature by:

(1) Undertaking analysis of multiple programmes and compound vulnerability at a finer (than county) scale of analysis. Many similar and more detailed individual level analysis must be undertaken if we are to truly understand how these programmes function from the equity perspective. Federal programmes can support this need by collecting applicant level information on vulnerability indicators including race, tenure, employment status, and other information enabling point level analysis of these critical pathways to equitable disaster recovery.

(2) Providing statistical analysis of the interactions (compounding) of flood specific vulnerabilities. This level of complex analysis requires multidisciplinary approach to analyse and address shortcomings in programmes affecting the most vulnerable populations. This type of work cannot be done by statistical analysis alone. Rather, a strong understanding of recovery programmes and associate policies, social science, geography, and demography is required when attempting such analyse at finer scales of geography.

Our results parallel those of other national and international flood recovery studies (Medd et al., 2015; Thrush, Burningham, & Fielding, 2005; Werritty, Houston, Ball, Tavendale, & Black, 2007) in identifying specific recovery challenges for different social groups, housing tenure situations, and age groups. The links between overal damage and disaster recovery funds identified here is undeniable and points towards programs that generally operate as intended. Although we found flood recovery assistance to be strongly associated with physical damage the relationships were more tenuous in places with higher social vulnerability. There are at least two potential confounding reasons for this: 1. Disaster recovery programs are not benefiting vulnerable populations, 2. Analyzing support data collected at the household level and demographic data collected at the tract level creates opportunity for ecological fallacy. Both of these challenges require more research aimed at identifying and understanding these relationships. However, the relationships between disaster outcomes and social vulnerability identified here provide unique opportunities for both better understanding how equity manifests in disaster and for developing more just recoveries in the future. More studies focused on the intersection of disaster, place, and people are needed if we (as a society) want to truly understand and influence future disaster outcomes in a positive manner.

Note

1. Full statistical results for this paper along with additional maps, charts, and graphs related to this work can be found at www.vulnerabilitymap.org.

Acknowledgements

We thank our colleagues from HORNE LLP who provided insight and expertise that greatly assisted the research.

Disclosure statement

No potential conflict of interest was reported by the authors.

Funding

This research was partially supported by funds from the U.S. National Science Foundation [grant numbers 1333190 and 1707947], under the Humans, Disasters, and the Built Environment programme.

ORCID

Christopher T. Emrich http://orcid.org/0000-0002-6773-7387
Eric Tate http://orcid.org/0000-0002-9587-3028
Sarah E. Larson http://orcid.org/0000-0002-9644-2019
Yao Zhou http://orcid.org/0000-0001-6422-7528

Data availability statement

The data that support the findings of this study are available from the corresponding author, Christopher T. Emrich, upon reasonable request.

References

Boyd, E., & Gonzales, O. R. (2011). *Community development block grant funds in disaster relief and recovery*. Washington, DC: Library of Congress, Congressional Research Service.

Burton, C. G. (2010). Social vulnerability and hurricane impact modeling. *Natural Hazards Review, 11*(2), 58–68.

Burton, C., & Cutter, S. L. (2008). Levee failures and social vulnerability in the Sacramento-San Joaquin Delta Area, California. *Natural Hazards Review, 9*(3), 136–149.

Collins, T. W., Grineski, S. E., & Chakraborty, J. (2018). Environmental injustice and flood risk: A conceptual model and case comparison of metropolitan Miami and Houston, USA. *Regional Environmental Change, 18*(2), 311–323.

Collins, T. W., Jimenez, A. M., & Grineski, S. E. (2013). Hispanic health disparities after a flood disaster: Results of a population-based survey of individuals experiencing home site damage in El Paso (Texas, USA). *Journal of Immigrant and Minority Health, 15*(2), 415–426.

Comes, T., Meesters, K., & Torjesen, S. (2019). Making sense of crises: The implications of information asymmetries for resilience and social justice in disaster-ridden communities. *Sustainable and Resilient Infrastructure, 4*(3), 1–13.

Cutter, S. L., Schumann III, R. L., & Emrich, C. T. (2014). Exposure, social vulnerability and recovery disparities in New Jersey after Hurricane Sandy. *Journal of Extreme Events, 1*(01), 1450002.

Doorn, N. (2017). Resilience indicators: Opportunities for including distributive justice concerns in disaster management. *Journal of Risk Research, 20*(6), 711–731.

Doorn, N., Gardoni, P., & Murphy, C. (2019). A multidisciplinary definition and evaluation of resilience: The role of social justice in defining resilience. *Sustainable and Resilient Infrastructure, 4*(3), 1–12.

Dunning, C. M., & Durden, S. (2007). *Theoretical underpinnings of the other social effects account*. Alexandria, VA: US Army Corps of Engineers Institute for Water Resources.

Elliott, J. R., & Pais, J. (2006). Race, class, and Hurricane Katrina: Social differences in human responses to disaster. *Social Science Research, 35*(2), 295–321.

Emrich, C. T., & Cutter, S. L. (2011). Social vulnerability to climate-sensitive hazards in the southern United States. *Weather, Climate, and Society, 3*(3), 193–208.

FEMA. (2016a). *National disaster recovery framework*. Washington, DC: Federal Emergency Management Agency.
FEMA. (2016b). *Individuals & households program unified guidance (IHPUG)*. Washington, DC: Federal Emergency Management Agency.
FEMA. (2016c). *Individuals and households applicant data, FIDA 25761 as of February 5, 2016*. South Carolina Disaster Recovery Office.
FEMA. (2018). *An affordability framework for the national flood insurance program*. Washington, DC: Federal Emergency Management Agency.
FEMA. (2019). *Fact sheet: Reasons why you might have been found ineligible by FEMA*. Retrieved from https://www.fema.gov/news-release/2019/03/13/fact-sheet-reasons-why-you-might-have-been-found-ineligible-fema
Finch, C., Emrich, C. T., & Cutter, S. L. (2010). Disaster disparities and differential recovery in New Orleans. *Population and Environment*, *31*(4), 179–202.
Fontenot, K., Semega, J., & Kollar, M. (2018). *Income and poverty in the United States: 2017*. Washington, DC: US Census Bureau. Government Printing Office.
Fothergill, A., & Peek, L. A. (2004). Poverty and disasters in the United States: A review of recent sociological findings. *Natural Hazards*, *32*(1), 89–110.
Gabe, T., Falk, G., & McCarty, M. (2005). *Hurricane Katrina: Social-demographic characteristics of impacted areas*. Washington, DC: Congressional Research Service, Congressional Information Service, Library of Congress.
GAO. (2018). *Federal disaster assistance: Individual assistance requests often granted but FEMA could better document factors considered*. Washington, DC: Government Accountability Office.
Green, R., Bates, L. K., & Smyth, A. (2007). Impediments to recovery in New Orleans' upper and lower ninth ward: One year after hurricane Katrina. *Disasters*, *31*(4), 311–335.
Grube, L. E., Fike, R., & Storr, V. H. (2018). Navigating disaster: An empirical study of federal assistance following hurricane sandy. *Eastern Economic Journal*, *44*(4), 576–593.
Hale, R. L., Flint, C. G., Jackson-Smith, D., & Endter-Wada, J. (2018). Social dimensions of urban flood experience, exposure, and concern. *JAWRA Journal of the American Water Resources Association*, *54* (5), 1137–1150.
Hamel, L., Wu, B., Brodie, M., Sim, S.-C., & Marks, E. (2017). An early assessment of hurricane Harvey's impact on vulnerable texans in the gulf coast region: Their voices and priorities to inform rebuilding efforts, Kaiser Family Foundation.
Heinz Center. (2000). *The hidden costs of coastal hazards: Implications for risk assessment and mitigation*. Washington, DC: Island Press.
Hewitt, K. (1997). *Regions of risk: A geographical introduction to disasters*. Singapore: Longman.
Ikeme, J. (2003). Equity, environmental justice and sustainability: Incomplete approaches in climate change politics. *Global Environmental Change*, *13*(3), 195–206.
Kamel, N. (2012). Social marginalisation, federal assistance and repopulation patterns in the New Orleans metropolitan area following Hurricane Katrina. *Urban Studies*, *49*(14), 3211–3231.
Kaufmann, M., Priest, S. J., & Leroy, P. (2018). The undebated issue of justice: Silent discourses in Dutch flood risk management. *Regional Environmental Change*, *18*(2), 325–337.
Koks, E. E., Jongman, B., Husby, T. G., & Botzen, W. J. (2015). Combining hazard, exposure and social vulnerability to provide lessons for flood risk management. *Environmental Science & Policy*, *47*, 42–52.
Kreiser, M., Mullins, M., & Nagel, J. (2018). *Federal disaster assistance response and recovery programs: Brief summaries*. Washington, DC: Government Accountability Office.
Kuehn, R. R. (2000). A taxonomy of environmental justice. *Environmental Law*, *30*, 10681–10703.
Li, W., Airriess, C. A., Chen, A. C.-C., Leong, K. J., & Keith, V. (2010). Katrina and migration: Evacuation and return by African Americans and Vietnamese Americans in an eastern New Orleans suburb. *The Professional Geographer*, *62*(1), 103–118.
Lindsay, B. R., & Reese, S. (2018). *FEMA and SBA disaster assistance for individuals and households: Application process, determinations, and appeals*. Washington, DC: Government Accountability Office.

Masozera, M., Bailey, M., & Kerchner, C. (2007). Distribution of impacts of natural disasters across income groups: A case study of New Orleans. *Ecological Economics, 63*(2–3), 299–306.

Medd, W., Deeming, H., Walker, G., Whittle, R., Mort, M., Twigger-Ross, C., ... Kashefi, E. (2015). The flood recovery gap: A real-time study of local recovery following the floods of June 2007 in Hull, North East England. *Journal of Flood Risk Management, 8*(4), 315–328.

Morrow, B. H. (1999). Identifying and mapping community vulnerability. *Disasters, 23*(1), 1–18.

Muñoz, C., & Tate, E. (2016). Unequal recovery? Federal resource distribution after a Midwest flood disaster. *International Journal of Environmental Research and Public Health, 13*(5), 507.

NASEM. (2019). *Framing the challenge of urban flooding in the United States*. Washington, DC: The National Academies Press.

NFIP. (2016). *NFIP SC claims Oct–Dec 2015*. National Flood Insurance Program.

NWS. (2016). *Service assessment: The historic South Carolina floods of October 1–5, 2015*. Maryland: U.S. Department of Commerce National Oceanic and Atmospheric Administration National Weather Service Silver Spring. Retrieved from https://www.weather.gov/media/publications/assessments/SCFlooding_072216_Signed_Final.pdf

O'Hare, P., & White, I. (2018). Beyond 'just' flood risk management: The potential for—and limits to—alleviating flood disadvantage. *Regional Environmental Change, 18*(2), 385–396.

O'Neill, E. (2018). Expanding the horizons of integrated flood risk management: A critical analysis from an Irish perspective. *International Journal of River Basin Management, 16*(1), 71–77.

Puente, S. (1999). *Social vulnerability to disaster in Mexico City*. Crucibles of Hazard: Mega-Cities and Disasters in Transition. J. Mitchell, United Nations University Press.

Reese, S. (2018). *Federal disaster recovery programs: Brief summaries*. Washington, DC: Congressional Research Service, Congressional Information Service, Library of Congress.

Rufat, S., Tate, E., Burton, C. G., & Maroof, A. S. (2015). Social vulnerability to floods: Review of case studies and implications for measurement. *International Journal of Disaster Risk Reduction, 14*, 470–486.

Rufat, S., Tate, E., Emrich, C. T., & Antolini, F. (2019). How valid are social vulnerability models? *Annals of the American Association of Geographers, 109*(4), 1131–1153.

Ryder, S. S. (2017). A bridge to challenging environmental inequality: Intersectionality, environmental justice, and disaster vulnerability. *Social Thought & Research, 34*, 85–115.

Sasanakul, I., Gassman, S. L., Pierce, C. E., Ovalle, W., Starcher, R., Gheibi, E., & Rahman, M. (2017). Dam failures from a 1000-year rainfall event in South Carolina. *Geotechnical Frontiers, GSP 278*, 114–124.

SBA. (2019). *SBA_home_applications_rpt_7405_SC-00031*. Washington, DC: Small Business Administration.

SCDRO. (2017). *South Carolina action plan for disaster recovery*. Columbia, SC: South Carolina Disaster Recovery Office, South Carolina Department of Commerce.

SCDRO. (2019). *SC 2015 flood cases*. Columbia, SC: South Carolina Disaster Recovery Office.

Schanze, J. (2006). *Flood risk management–a basic framework. Flood risk management: Hazards, vulnerability and mitigation measures* (pp. 1–20). Dordrecht: Springer.

Siders, A. (2019). Social justice implications of US managed retreat buyout programs. *Climatic Change, 152*(2), 239–257.

Smith, G. P., & Wenger, D. (2007). Sustainable disaster recovery: Operationalizing an existing agenda. In H. Rodriguez, E. L. Quarantelli, & R. R. Dynes (Eds.), *Handbook of disaster research* (pp. 234–257). New York, NY: Springer.

Spielman, S. E., Folch, D., & Nagle, N. (2014). Patterns and causes of uncertainty in the American Community Survey. *Applied Geography, 46*(0), 147–157.

Thaler, T., Fuchs, S., Priest, S., & Doorn, N. (2018). Social justice in the context of adaptation to climate change—reflecting on different policy approaches to distribute and allocate flood risk management. *Regional Environmental Change, 18*(2), 305–309.

Thaler, T., & Hartmann, T. (2016). Justice and flood risk management: Reflecting on different approaches to distribute and allocate flood risk management in Europe. *Natural Hazards, 83*(1), 129–147.

Thomas, D. S., Phillips, B. D., Lovekamp, W. E., & Fothergill, A. (2013). *Social vulnerability to disasters*. Boca Raton, FL: CRC Press.

Thrush, D., Burningham, K., & Fielding, J. (2005). *Vulnerability with regard to flood warning and flood event: A review of the literature*. Environment Agency R&D Report W5c-018/1, Environment Agency, Bristol, UK.

Tierney, K., & Oliver-Smith, A. (2012). Social dimensions of disaster recovery. *International Journal of Mass Emergencies & Disasters, 30*(2), 123–146.

University of Maryland, Center for Disaster Resilience, and Texas A&M University, Galveston Campus, Center for Texas Beaches and Shores. (2018). *The growing threat of urban flooding: A national challenge*. College Park: A. James Clark School of Engineering.

Vinik, D. (2018). *'People just give up': Low-income hurricane victims slam federal relief programs*. Politico. Retrieved from https://www.politico.com/story/2018/05/29/houston-hurricane-harvey-fema-597912.

VMAP. (2018). *Social vulnerability mapping tool*. The Vulnerability Mapping and Analysis Platform. Retrieved from https://www.vulnerabilitymap.org/

VMAP. (2019). *FEMA payouts updated for 2017*. The Vulnerability Mapping and Analysis Platform. Retrieved from https://www.vulnerabilitymap.org/

Walker, G., & Burningham, K. (2011). Flood risk, vulnerability and environmental justice: Evidence and evaluation of inequality in a UK context. *Critical Social Policy, 31*(2), 216–240.

Werritty, A., Houston, D., Ball, T., Tavendale, A., & Black, A. (2007). *Exploring the social impacts of flood risk and flooding in Scotland*.

Wisner, B., Blaikie, P., Cannon, T., & Davis, I. (2004). *At risk: Natural hazards, people's vulnerability and disasters*. New York, NY: Routledge.

Approaches to state flood recovery funding in Visegrad Group Countries

Lenka Slavíková, Pavel Raška, Kazimierz Banasik, Marton Barta, Andras Kis, Silvia Kohnová, Piotr Matczak and Ján Szolgay

ABSTRACT

Flood recovery is an important period in the flood risk management cycle. Recently, flood recovery has become viewed as an opportunity for future flood damage mitigation. Financial flows to cover flood damages and rules regarding their allocation are crucial for supporting or undermining mitigation efforts. In this paper, we map and compare state flood recovery funding in the so-called Visegrad Group Countries (V4), i.e. Czechia, Hungary, Poland and Slovakia, over the past 30 years of their democratic history. We apply a qualitative comparative approach to identify differences and similarities in risk sharing and state flood recovery funding approaches among these countries. Additionally, we reveal how risk sharing is addressed by existing flood recovery funding schemes. The results indicate that national governments have a low willingness to institutionalise *ex-ante* compensation schemes. *Ad hoc* instruments initiated shortly after disastrous flooding usually do not provide incentives to reduce future flood damages.

1. Introduction

Recovery is among the key activities in flood risk management (FRM); it comes after the immediate threat passes, a focus on reconstruction takes place and the affected area attempts to return to 'normality' (Medd et al., 2015). While the importance of recovery for future damage reduction used to be underestimated and recovery was considered only a temporary limited period of the so-called disaster cycle (Drennan, McGowan, & Tiernan, 2016), recent studies have increasingly acknowledged that it has a much

Supplemental data for this article can be accessed https://doi.org/10.1080/17477891.2019.1667749

greater potential to reduce vulnerability (Medd et al., 2015; Moatty & Vinet, 2016). A post-disaster situation opens a 'window of opportunity' for more transformative changes in complex systems (Brundiers & Hallie, 2018; Thomalla et al., 2018). Recovery funding schemes play an essential role here – they can either encourage people to re-build in the same manner in the same places, or they can provide incentives for more considerate land use policies and changes in behavioural adaptation to reduce future flood damage (Aerts et al., 2018).

Despite the current paradigm shift towards multilevel flood risk management (Thaler & Priest, 2014), national or federal governments play a significant role in FRM: 'They shape regulatory frameworks, bear the ultimate responsibility for crisis management and often provide a critical amount of money for post-disaster emergency relief and flood prevention measures' (Slavíková, 2018, p. 1). Financial schemes are among the key governmental approaches to post-flood recovery. A common feature of recovery financing is the provision of resources shortly after a flood, but the particular schemes may differ in terms of resource regeneration and allocation rules. There has been no agreement upon the optimal design and extent of state recovery funding, as the contextual factors matter greatly (see the discussion on justice in Thaler & Hartmann, 2016; or Penning-Rowsell & Priest, 2015). It is recommended, however, that financial schemes aim at flood risk reduction by linking recovery strategies with prevention and mitigation strategies (Suykens, Priest, van Doorn-Hoekveld, Thuillier, & van Rijswick, 2016).

It has been indicated that in countries where central governments have traditionally played a strong role (such as in post-socialist European countries), state flood recovery funding tends to crowd out individual efforts (Raschky, Schwarze, Schwindt, & Zahn, 2013; Slavíková, 2018). However, robust comparative evidence of the risk-sharing approaches and recovery scheme designs in this part of Europe is missing. On the other hand, despite some contextual similarities, the individual post-socialist states have now experienced three decades of a rather independent democratic evolution. Therefore, their approaches to state recovery funding might differ and create variable conditions for implementing international FRM strategies (EC, 2007/60/EC).

With ongoing eastern enlargement of the EU in the last two decades, the FRM strategies are being transferred to new institutional and environmental contexts. This raises a question about possible implementation barriers that may hamper the effort to reconcile the EU FRM strategies (see e.g. Slavíková, Raška, & Kopáček, 2018). The goal of this paper is to provide a comparative perspective on state flood recovery[1] funding for private damages in the so-called Visegrad Group Countries, i.e. Czechia, Hungary, Poland and Slovakia. In particular, we address two questions: (i) How (if so) is risk reduction addressed by existing flood recovery funding schemes? and (ii) What are the differences and similarities in risk-sharing and state flood recovery funding approaches among these countries? The comparison covers the period of three decades of the recent democratic history of these four countries. In this respect, the paper builds on the research by Suykens et al. (2016) and Priest, Penning-Rowsell, and Suykens (2016) to extend the empirical evidence and to further identify research issues relevant to research on risk sharing at the country level. The research is performed by a comparative analysis of legislation, strategies and policies using qualitatively-based desk research and a literature review.

The paper is divided into three main sections. First, we review the (changing) approaches to state flood recovery funding with a focus on risk sharing and the incorporation of mitigation incentives. Then, we describe the data collection and research design and introduce four key findings. The subsequent discussion explains the situation in the Visegrad Group Countries within the broader context of European Union (EU) integration.

2. Recent approaches to state flood recovery funding

Over the past two decades, the recovery period has become viewed as an opportunity for FRM. According to Moatty and Vinet (2016, p. 2), '... it is a time when local stakeholders gather around a common goal: managing the recovery of society. The damage and destruction open the possibility of rebuilding otherwise.' This view is in contrast with a previous understanding of recovery that stressed early restoration as the main goal (Thomalla et al., 2018). The remaining question is what the rebuilding of affected places should look like and who the main actors of possible change are. Financial flows to cover flood damages and the rules of their allocation are crucial for supporting or undermining this effort. In the following review, we focus on the responsibility of the state in recovery funding, its critique and on innovative state-based financial schemes addressing risk reduction.

In a majority of countries, national and/or federal governments share the responsibility for flood damages (Priest et al., 2016). This is based on existing legislation, *ad hoc* decision making and/or customary arrangements. In particular, states are obliged to restore state/public property, such as infrastructure, but they also tend to compensate for private flood damages (Suykens et al., 2016). However, when insufficiently coordinated or balanced with private-based recovery funding (e.g. insurance), state efforts are subject to increasing criticism. State recovery funding schemes shift costs from affected people to the wider taxpaying population (Sandink, Kovacs, Oulahen, & Shrubsole, 2016). If they are implemented in parallel with private flood insurance, they might crowd out the insurance demand (Raschky et al., 2013) or generally limit incentives for property-level mitigation efforts (Sandink et al., 2016). According to Davies (2016), reliance on state flood recovery funding accelerates the moral hazard situation.

Following this critique, different pathways for responsibility-sharing changes in FRM have been drafted. First, a more institutionalised sharing of the responsibility for flood recovery and mitigation with local governing bodies and individuals has been suggested (Thaler & Priest, 2014). This 'community-based' recovery is supposed to raise the higher flood risk awareness of local actors (Henstra, Thistlethwaite, Brown, & Scott, 2019; Meijerink & Dicke, 2008). Second, pre-disaster recovery planning should be activated at all governance levels to clearly delineate the priorities during the recovery period and how they are to be (or not) funded; e.g. the preparation of adjustments to land-use plans to speed up resettlements (Moatty & Vinet, 2016; Mohd, Abu Bakar, Hassan, & Hussain, 2017).

Essentially, the integration of all risk reduction components into the core functions of recovery funding schemes is advocated. Priest et al. (2016) mention the following key incentives relevant for state recovery funding: (a) exclusion of property from coverage; (b) application of minimum standards or special conditions imposed on a property; (c) limitation of the level of indemnification or the amount compensated; and (d) standard retrofitting of buildings following flood events. The practical implementation of these components has been described by numerous authors in many contexts, in which either co-financed schemes or the role of recovery in flood risk reduction is stressed. For example, in Belgium, the government reduces compensation or even excludes (newly built) property from coverage according to existing risks (Suykens et al., 2016). The Netherlands and Iceland require co-financing of flood damages by affected households (Priest et al., 2016). The Canadian government declares that its financial aid will not cover losses that are reasonable and affordably insurable (Sandink et al., 2016). In contrast, positive incentives might be given as well. In Canada, from 2008, additional 15% federal disaster relief might be provided if mitigation measures are applied at the household level (Sandink et al., 2016). Similarly, in Australia, since 2007, the principle of 'betterments' has been incorporated into government-funded disaster relief – i.e. if a state redevelops damaged public infrastructure to a more resilient standard, it will gain extra resources from the federal budget (Drennan et al., 2016). In the United States, the recent focus has been on federally supported programmes for voluntary property acquisition (such as the Hazard Mitigation Grant Program), through which owners of severely damaged properties are bought out from floodplains (Tate, Strong, Kraus, & Xiong, 2016). All of these innovative flood recovery schemes and adjustments aim at building the missing linkage between recovery and flood risk reduction.

In the existing scientific literature, there is only fragmented knowledge on how these developments in state recovery funding are reflected in the four post-socialist states that we focus on. The best evidence available is for *Poland*; according to Priest et al. (2016), Poland relies on aid to victims via a disaster fund and *ad hoc* solutions, such as tax exemptions, individual market-based insurance and *ad hoc* government compensation. Flood risk reduction components are, however, missing in these solutions. In the *Czech* context, Slavíková (2018) deals with general responsibility sharing for flood damages – the central government pursues *ad hoc* compensation without addressing risk reduction. Raška and Dostál (2017) showed that although the recent change in the disaster recovery scheme was clearly influenced by post-socialist political democratisation, it has maintained the pivotal role of the national government. In *Hungary*, special attention is given to flood storage compensation schemes for agricultural land (Weikard, Kis, & Ungvári, 2017), but scientific literature on state recovery funding after flood disasters is generally absent. Only scarce information has been published on *Slovakia*, where authors focus mainly on private flood insurance (see, e.g. Solín, Madajová, & Skubinčan, 2015). The abovementioned lack of systemic research is the reason why more robust comparative evidence of state flood recovery funding in Czechia, Hungary, Poland and Slovakia is needed.

3. Research design and data collection

The Visegrad Group Countries (also called V4) are represented by four Central European post-socialist states that have shared important moments in history, geopolitics and culture (Wandycz, 1992). The group was formed in 1991 in the Hungarian town of Visegrad through presidential signatures of the declaration (Declaration, 1991). In subsequent years, the V4 coordinated their EU and NATO accession efforts and aimed to enhance coordination in the Central European region in many fields. The above noted similarities of the V4 countries have produced a common background for current efforts in national policies and the level of their success (see, e.g. Hruška, Czapiewski, & Kovács, 2015; Ivanová & Masárová, 2018 among many others). On the other hand, the environmental variability across the V4 countries, which share only a few significant transboundary river basins, has resulted in low attention being paid to comparative perspective on flood risk reduction strategies in the academic literature.

Our research aims to compare state flood recovery funding approaches and is based on an external comparative analysis of legal, strategic and policy documents (Suykens et al., 2016). The analysis of the documents was conducted by co-authors (national experts) from each country and coordinated in several steps (Figure 1). The national experts have been chosen to conform with the following criteria. While they were naturally supposed to have a broad expertise in FRM, our aim was to establish an interdisciplinary team that finally included institutional economists, geographers, hydrologists and sociologists. Although it may be argued that group of economists would have the most apt expertise for the research aims, our decision allowed for validation of the research question and answers from different perspectives; thus creating more coherent image of the recovery funding schemes. Second, two national experts from each country have been involved to facilitate within-country discussions.

First, based on a review of the existing literature (see Section 1) on flood recovery schemes, the sample structure of a data collection template was prepared (Supplement 1). Each item of the template allowed for an open narrative answer. The open questions enabled us to outline the diverse realities of the individual countries, whereas the narrative approach allowed us to trace and comprehend the causal logic of the national case studies (cf. George & Bennett, 2005) regarding the evolution of particular financial schemes. Then, the filled version of the template using the example of Czechia was prepared.

Second, the template was distributed to the co-authors – national experts (two for each V4 country) who were asked to fill in the information for their country and provide any further comments regarding the formulation of the questions and character of the required data. This proved to be crucial, as the realities of the countries differ and contain specific FRM mechanisms that are not present in other countries. The national experts were asked to analyse the FRM-related legislation over the past decades together with country-wide strategies and policies. In addition, they were asked to provide references to the literature (also in their national language) that addresses the issue of financial recovery schemes and the evaluation of their efficiency and effectiveness.

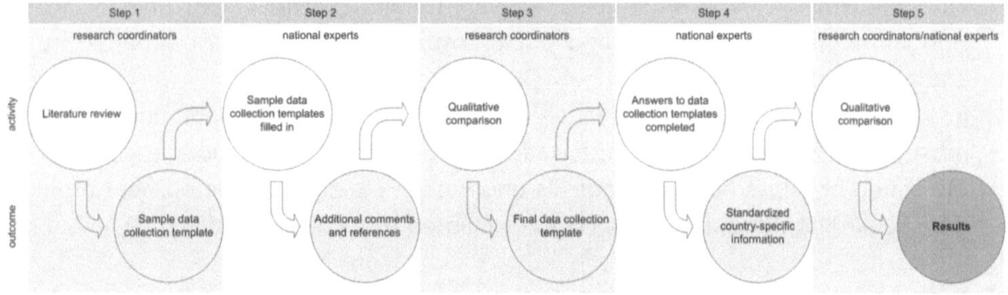

Figure 1. Scheme of the data collection and analysis. Source: authors.

Third, the obtained answers were analysed using a qualitative comparative approach (Young et al., 2006). The results, along with the comments from co-authors, were used to refine the data collection template items and improve their explanation to address variable FRM situations in the V4 countries. The refined template with explanations was re-sent to the national experts.

Fourth, the national experts completed the unclear parts and realigned the information with the other countries. Using this approach, it was possible to obtain standardised country-specific information.

Finally, (fifth), the collected templates were re-examined with the qualitative analyses and organised into a comparative perspective; the results were then once again checked and refined by national experts. An overview of key legislation in each country was developed and forms a Supplement 2.

4. Results

Results that emerged from the qualitative comparison of the data templates show the differences and similarities in risk sharing and state flood recovery funding among the V4 group and address risk reduction components in existing recovery funding schemes. The comparative analysis revealed that the respective countries share many similar features in terms of risk sharing, the design of existing flood recovery schemes and national state involvement. In most of the respects addressed, however, at least one of the countries acts differently from the others (for details, see below). Reflecting on the research questions, the similarities and deviations that will be further described in the following sub-sections are summarised below:

(i) Despite repetitive disastrous flooding (from the late 1990s until now), there are very low dynamics in state recovery funding scheme development. The prevailing national government approach is to provide relief on an *ad hoc* basis and to use the broader *tax payers' money* to cover public and partly private flood damages (the exceptions are the Hungarian state-based insurance schemes for farmers and households).

(ii) The explicit *declaration of responsibility sharing* for flood damages among the state, regional units, and citizens is to a large extent missing. National governments are expected to contribute to individual flood damages. The responsibility of citizens under threat is rather blurred (except for the developed comprehensive flood strategy in Czechia).

(iii) Individual *market-based flood insurance* exists in all countries and functions rather independently from the state flood recovery funding provision (only in the case of Hungarian farmers do private and public insurance schemes exist simultaneously and are coordinated).

(iv) *Risk reduction incentives* are not widely applied in state flood recovery funding. States mostly limit the amount of compensation per household due to budget constraints. Mitigation of future flood damages is undertaken mainly via development prohibitions regarding floodplains, not via rules tailored to recovery funding.

Four themes listed above reflects the topics explored though the circulated templates (Supplement 1).

4.1. *Low developmental dynamics and the preference for* ad hoc *solutions*

In the democratic history of the V4 (i.e. from 1989), the first disastrous flooding occurred in 1997 (Oder and Danube river basin) or 1998 (Tisza river basin). In the subsequent 15 years, large fluvial flooding continued to come every 3–4 years, causing repetitive damage. Since 2013, there has been a relative flood silence (except for Slovakia, which has experienced disastrous flash floods and snow-melting floods almost every year).

Before the first large flooding in the post-socialist development of the V4 countries, responsibility sharing and the recovery-resilience nexus in FRM had not been on the governmental agenda. These issues arose just after the first disastrous floods in 1997–1998, after which all countries focused on developing new crisis management codes and establishing warning systems. In addition, large flood protection measures have been planned and funded. In Hungary, a new flood defence strategy called for the development of a series of emergency reservoirs along the river Tisza. In Czechia in 2000, a comprehensive strategy for flood protection (Czech Flood Protection Strategy, 2000) was adopted, and its implementation has been evaluated annually. In Poland, a crisis management system was established after the 1997 flood, and the large flood protection programme on the Odra River was executed based on a loan from the World Bank.

In terms of state flood recovery funding, all V4 countries have applied *ad hoc* compensations for a part of private damages. The portion of these damages covered is difficult to assess and varies among the countries – from 30% after the floods in 1997 and 2002 in Czechia (Slavíková, 2018), to approx. 30% in Slovakia, to 50% after the floods in 1997 in Poland (Linnerooth-Bayer & Amendola, 2003). In Hungary, Vari, Linnerooth-Bayer, and Ferencz (2003, p. 591) have indicated that at the beginning of the millennium, the

> government has been to compensate victims for 100% of their losses if the cause of the flood is insufficiency of the levees, but for only 50% of their losses if the cause is not directly connected to the government's negligence in providing protection.

While in all countries the recovery schemes are designed to provide support to individuals and farmers, in Czechia and Slovakia, recovery funding may also be claimed by businesses. Providing advantageous loans for the reconstructions of households was also an option in Czechia after the floods in 2002 and 2006.

The prevailing compensation approach combined with repeated disastrous flooding has created pressure on the state budgets. Financial sources for these compensations were released mainly from state budget reserves (i.e. general tax payers' money). After the first large flood in 1997, the Czech state issued flood bonds that could also be bought by inhabitants. In 2010, the Czech government decided that the people will contribute a special 'flood income tax' for one year to cover the flood damages. Approximately 160 mill. EUR was generated in this way (Kučera, 2012). In Slovakia and Hungary, the governments started to create budget reserves for unpredictable events that might also be used for *ad hoc* flood damage compensations. In Poland, after the 1997 flood, the government created several tax exemption schemes for the flood victims. Moreover, the specific budget reserve is held by the Department of Disaster Relief of the Ministry of Internal Affairs. The size of the reserve depends on the capacity of the state budget and on the losses caused by natural disasters in previous years. Reserve funds are mainly spent on the reconstruction of infrastructure damaged as a result of natural disasters and on help to victims. Based on the Polish Crisis Management Act (2007), Polish municipalities are obliged to create a budget reserve of 0.5–1% of their revenue to cover disaster damages.

After 2002, the newly established EU Solidarity fund repeatedly contributed to national states to cover flood damages. In Czechia after the floods in 2002, the Solidarity covered approx. 7% of the total damages (Slavíková, 2018), but the Fund basically does not provide compensation for private losses.

Based on the description above, the customary practice of the national governments to significantly contribute to private flood damages on an *ad hoc* basis is apparent for the entire V4. From the late 1990s until recently, it has been difficult to observe any significant changes in compensation strategies in terms of state declarations (see further text), resource generation and allocations. We denote this situation as a low developmental dynamics of the recovery funding schemes. After a large flooding, national governments fulfil their responsibility to help affected inhabitants, and via administrative procedures, they allocate money from state budget reserves.

The only exception to this is *Hungary*, where the high level of solidarity in face of repeated damage and public budget constraints has resulted in the institutionalisation of two instruments. State recovery funding schemes have been developed for (a) farmers and (b) households in high risk locations:

(a) In 2011, the *compulsory public insurance scheme* for farmers was established (the Hungarian Natural Disaster Act, 2011). Farmers above a certain size limit who also apply for EU-funded agricultural subsidies have been obliged to contribute to a state-funded disaster damage compensation scheme. The size limit depends on land use and varies between 1 hectare (orchards) and 10 hectares (regular crops). Farmers that are below the size limit can choose to take part in the scheme and thereby become covered by insurance. When a natural disaster comes, compensation from the state covers 50% of the actual damage. However, farmers that also have market-based insurance can be eligible for a higher percentage, up to 80% covered by the public and private schemes together. If in a given year the public insurance fund is not sufficient for covering the eligible damages, the deficit is covered by the state budget.

(b) In 2003, the Hungarian government created a state-sponsored insurance fund called the *Wesselenyi Fund* that was available only to households in flood risk locations where market-based insurance was difficult/costly to obtain. Households that decided to join the scheme paid an annual insurance fee and became eligible for compensation in case of being flooded. The scheme, however, never truly gained traction. At the end of 2016, it comprised less than 200 households, and the national government decided to abolish it. Households that were already members could have continued paying the annual insurance fee and kept the government-assured help, but new members were no longer allowed to join.

4.2. The need for better declaration of responsibility sharing?

None of the V4 countries has codified an explicit diversification of responsibilities for flood recovery among the government, other stakeholders, and affected individuals and companies. Generally, there are persisting expectations of strong governmental involvement in flood recovery schemes (Raška, 2015).

In particular, in *Hungary*, the national government has no formal obligation to compensate for private damages after a flood, but it frequently does (see Vari et al., 2003, quoted in section 3.1). In turn, the absence of declared government responsibility indicates that households should protect themselves directly or via market-based insurance (to be ready to cover flood damages). On the other hand, the customary evolution has provided disincentives to individual actions. The gap between these two approaches has not been widely discussed in Hungary. The legally binding responsibility sharing has been developed only for farmers who are obliged to contribute to the compulsory public insurance scheme. Similarly, in *Slovakia*, the state and the public administrations are not responsible for covering private flood damages. In decentralised flood risk management plans, the initiative and active involvement of inhabitants are expected. The Slovakian national government may and often does decide to compensate for private flood damages.

In *Poland*, the National Water Management Authority and the national and local governments have a general responsibility for flood protection that also includes flood recovery (Polish Water Act, 2017, Art. 163, Item 6). They should ensure the return of the territory to the state that it was in before a flood. The governmental disaster fund establishment supports this general declaration, although actual aid and the specific conditions of help vary on a yearly basis. Nevertheless, municipalities can apply for aid in the case of a natural disaster (not only a flood). Furthermore, based on the Polish Act on Social Assistance (2004), the municipalities are to financially assist people affected by disasters. This arrangement imposes generally declared duties on the state and other public bodies. The responsibility of private subjects for flood damages (households, entrepreneurs, farmers) are not specified, however.

In *Czechia* in 2000, the comprehensive governmental document Czech Flood Protection Strategy (2000) was developed. One of its declared goals was 'to define the scope of responsibilities in the flood protection system at levels of the State and local administration, and responsibilities of public (individuals) and business companies' (p. 4). In terms of flood recovery, it is stated that 'it is everybody's duty to protect adequately his property against floods' (Czech Flood Protection Strategy, 2002, p. 9). The document also states that there is market-based insurance widely available, so having the private property insured should be the pursued strategy. The option of state financial involvement in recovery is acknowledged, but 'the State financial assistance for the housing restoration should preferably be implemented through loans and credits and this system could offer some advantages to those who have concluded insurance agreements' (Czech Flood Protection Strategy, 2002, pp. 9–10). In fact, as analysed by Slavíková (2018), the adoption of the strategy brought only minor changes to real recovery funding when comparing the situation after the floods in 1997 and 2002. Therefore, the presence of a responsibility sharing declaration does not need to shape real decision making. Furthermore, the Czech Recovery Funding Law (2002) declares: 'if there is a disaster classified as crisis, the state might provide compensations via particular ministries to municipalities, people and entrepreneurs.' The goal is to bring the affected territory back to normal. Although no particular financial schemes have been established, the active role of the national governments in financing is acknowledged.

An overview of the legislation and other key governmental documents categorised according to the V4 countries is provided in the Supplementary material.

4.3. Coordination with existing market-based insurance schemes

All V4 countries have market-based insurance covering flood damages provided by multiple insurance companies. Insurance penetration increases with growth in the mortgage market. In Czechia, Slovakia and Poland, insurers diversify premiums according to risks – i.e. high-risk territories are more difficult or impossible to insure. The zones with different levels of risk are defined by private insurers (e.g. Association of Insurers in Czechia), however, and do not perfectly coincide with flood hazard and risk zones as defined by national and regional authorities. In Hungary, market-based insurance is not offered for buildings in floodplains where flood protection (dikes) does not exist, but it is available everywhere else under the same conditions.

In most countries, market-based insurance systems have developed somewhat independently from complex FRM strategies. Coordination with state compensation policies is largely missing. In *Poland*, after the floods in 1997, there was an attempt to build a universal flood insurance system with state involvement. This attempt met with criticism and failed, as the public was apparently resistant to new 'taxation' schemes. In *Czechia*, after the floods in 2002, there was a political proposal to make insured households contribute to other flood damages. After a great disagreement with the insurance industry, this proposal was cancelled.

Only in *Hungary* might farmers contributing to the compulsory public insurance scheme obtain higher total compensation if they are privately insured (80% of damages instead of 50%). The private agricultural damage insurance scheme also enjoys a state subsidy, making these insurance products more affordable to farmers. Furthermore, the Hungarian government provides the post-flood relief only to households that are willing to sign market-based insurance contract to cover future damages.

4.4. (Missing) risk reduction incentives

In the literature review (Section 1), we showed that in many countries worldwide, risk reduction incentives have been incorporated into state recovery funding schemes. Via these incentives, repeated flood damage could be reduced. Within the V4, the states have used different strategies, mainly to limit the amount of flood damage compensated per household. As most of these schemes were set up by government decision just after a flood, we can refer mainly to past customary practices.

In *Slovakia*, each household estimated the extent of the flood damage. This estimation was verified by an expert commission, and a certain percentage of damage was compensated with no risk mitigation requirements. After a disastrous flood in 2010, the compensation varied between 300–1300 EUR/household (Solín et al., 2015). In *Czechia* and *Poland*, the calculation of flood damage was not required, and the flat rates per household were set for different situations and purposes – e.g. a Polish household received 1500 EUR for general aid, 4500 EUR for damage building renovation and 25,000 EUR for a complete building reconstruction. After the 1997 floods, Czech households of demolished houses received approx. 6000 EUR – they only needed to prove ownership and that the property was earmarked for demolition due to flood damage. After disastrous floods in 2002, the Czech Governmental Direction 396/2002 enabled individuals to apply for state loans of up to 32,000 EUR to build or buy a new house or flat, provided that it would not be located on the floodplain. After the floods in 2013, Czech farmers were able to claim the actual flood damage compensation for crops, animals and storage. Damages were calculated with the use of governmental prescriptions and approx. 1/3 to 1/2 of them were compensated.

The situation differed slightly in *Hungary*. Partly damaged houses that were still suitable for living were excluded from state compensation – the emergency relief for these households was assured in cooperation with the Red Cross and other NGOs. Severely damaged houses that were not possible to rebuild were compensated for by the national government, but the provision of the state aid was conditional on entering into a new flood insurance contract. The declared goal of the Hungarian government was to reduce its future compensation needs. After the floods in 2010, the government compensation for severely damaged houses was approx. 30,000 EUR at maximum.

Based on this overview, it is clear that the V4 governments are rather reluctant to precondition state flood recovery money distribution, e.g. with special (re)construction conditions or standard retrofitting. We identified only two cases (a private insurance contract requirement in Hungary and a relocation condition in the case of Czech loans) in which national governments required certain risk reduction actions. Otherwise, risk reduction is undertaken mainly via development restrictions or prohibitions that apply mainly to new constructions.

5. Discussion

The comparative evidence of state flood recovery funding in V4 countries has revealed that there is generally a low willingness among national governments to institutionalise *ex-ante* complex instruments with clear resource generation, allocation and coordination rules. Despite the evidence showing that general statements on responsibility sharing exist in some of the V4 countries, as do certain mechanisms for financial recovery, in most cases, state recovery funding is not obligatory, uses an *ad hoc* generation of sources, is only scarcely coordinated with other financial instruments, and does not provide incentives for decreasing future flood risk. In this respect, it might be concluded that risk reduction is only scarcely addressed by existing flood recovery funding schemes in the V4 countries. A detailed review of national policies, strategies and legislation provides a further varied picture that outlines the different pathways for state flood recovery funding in the V4 countries (see Figure 2). This allows us to address the differences and similarities in risk sharing and state flood recovery funding approaches among the V4 countries. In addition, the comparative cross-country perspective enables us to discuss the key preconditions and drivers of the current situation.

First, there is a question of the factors that cause the low dynamics of the state flood recovery schemes. Basically, in all the V4 countries, the first yet modest attempts to declare responsibility sharing and to accentuate prevention measures within FRM date back to the disastrous floods at the turn of millennium (in 1997–2002). However, these efforts did not establish clear mechanisms for responsibility sharing, nor did they provide instruments and incentives for connecting financial recovery with future flood risk reduction. It might be presumed that in the transition period from a command and control system to a democratic market-based economy, the V4 governments pursued

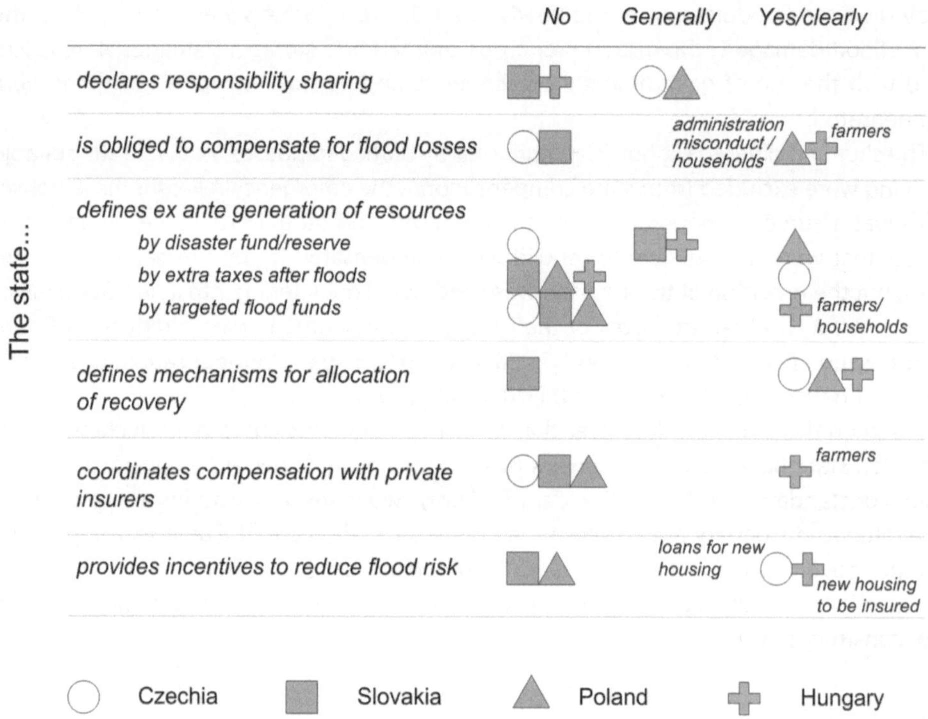

Figure 2. Comparative summary of state financial recovery efforts in V4 countries. Source: authors.

other demanding agendas. The FRM was brought into focus only after the first disastrous floods clearly proved that former instruments are inefficient (Raška & Dostál, 2017, see also Ženka, Pavlík, & Slach, 2017 for broader context). In the period after 2000, the EU Floods Directive (EC, 2007/60/EC) may be perceived as a catalyst of changes in national FRM policies. The obligation to transpose its goals and approaches to national legislation resulted in a broad debate on policy and among practitioners and has placed various aspects of FRM on the agenda.

Second, there have been relatively low efforts made to design *ex-ante* state flood recovery schemes that would support future flood risk reduction in the V4 countries. In this respect, the EU Floods Directive (EC, 2007/60/EC) also did not provide sufficient incentives, as its outcomes are generally understood to support planning documents with flood hazard and risk maps and flood risk management plans, while state flood recovery schemes are left to national policies. The efficiency of flood risk reduction is dependent, however, on the complementarity of various instruments. In the V4 region, where spatial planning regulations are considered the main approach to reduce future flood losses for private properties, there is a low willingness to design *ex-ante* state flood recovery schemes using financial instruments. Such a situation may be ascribed to unclear roles and expectations of the different entities within FRM. It has been previously shown that in the V4 region, direct regulation to achieve political goals is favoured compared with the application of financial instruments, as the public better understands and is more willing to accept strict regulations than flexible solutions (see, e.g. Istvan, 2015; Lemkowska, 2019; Slavíková, Vojáček, & Smejkal, 2017). In addition, if any state flood recovery schemes exist (e.g. Hungary, Czechia), they are mostly disconnected from FRM strategies. The persisting public perception of floods as temporary discrete events may be considered one of the drivers of such situations (Raška, 2015). This calls for further efforts to link the recovery schemes with individual-level flood mitigation and prevention, e.g. via various types of disincentives for the (re)development of flood-prone areas (Suykens et al., 2016).

Third, there is hardly any consensus as to what extent and how (or even if) people should be held responsible for their private flood damages. There is currently no lively discussion on this issue, and market-based insurance is considered the main financial instrument for individuals. In most cases, the national experts in our research stated that the only declaration of responsibility sharing is linked to the transposition of the EU Floods Directive (EC, 2007/60/EC) to national legislations. The contribution of the directive to better responsibility sharing and participatory approaches to FRM is, however, still subject to debate (cf. Newig, Challies, Jager, & Kochskämper, 2014). Additionally, the insight into the national legislation and FRM instruments in the V4 countries indicated that the drafted participative approach has a rather declaratory role without clear implications for individuals and governments. In the V4 region, national governments are still expected to compensate as much as they can. This situation causes mismatches and inefficiencies that are acknowledged by scientists and experts. According to Solín et al. (2015, p. 184), for example, 'the government should proceed with the systematic, legislative solution that would accurately specify the role and instruments of the state, local authorities and private insurance companies in the strategy mitigating the negative effects of floods.' The authors referred to the Slovakian situation in particular, but their request could also be applied to other V4 countries. There are also indications that government involvement crowds out individual motivation to contribute to flood risk reduction, as previously described by Raschky et al. (2013). This suspicion is confirmed, e.g. by the evolution of the Hungarian Wesselenyi Fund, which offered an obligatory state-guaranteed insurance for people living in high-risk locations but gained almost no traction.

6. Conclusions

The paper aimed to provide a comparative perspective on state flood recovery funding for private damages in the V4 group, which includes post-socialist countries that have shared certain similar economic, political and institutional developments over the past three decades. The key findings of the qualitative analysis might be summarised as follows:

(i) In the past 30 years, there have been relatively low efforts to design *ex-ante* state flood recovery funding schemes, national governments widely compensate private damages on *ad hoc* bases after extreme floods.
(ii) There is hardly a consensus as to what extent and how (or even if) people should be held responsible for flood damages. The topic has been on the national government agendas only scarcely.
(iii) The market-based flood insurance that exists in all the V4 countries has developed independently and is mostly not coordinated with the state flood damage compensation policies.

(iv) Existing state flood recovery funding schemes provide only limited flood risk reduction incentives. Prohibitions based on spatial plans are the preferred instruments for reducing flood plain development and flood damage to private properties.

(v) The EU Floods Directive (EC, 2007/60/EC) and its transposition to national legislations and policies brought FRM strategies into public debates, but its contribution to flood risk reduction via recovery funding is limited.

Although deviations from these general statements have been discussed, it can be concluded that the V4 countries have not widely applied the current trend to use state flood recovery funding schemes strategically. This situation may be described as a drift, where the communist regime legacy appears in the form of the popular reliance on the state that is expected to bear the responsibility for flood recovery. As a result, recovery funding rather lean towards reactive and aid-oriented approach. Future research efforts should therefore focus on political and public perceptions of and willingness to adopt specific funding instruments (such as premiums, betterments, or reductions and exclusions) in the V4 region.

Note

1. For the purposes of this research, we exclude financial schemes that are focused on immediate threat mitigation during crisis situations (e.g. covering expenses for food provision or the loss of salaries during reconstruction works).

Acknowledgements

This article is based upon work from COST Action LAND4FLOOD (CA16209) supported by COST (European Cooperation in Science and Technology), www.cost.eu. L.S. and P.R. would also like to thank the Operational Programme Research, Development and Education of the Czech Republic for financing the project Smart City – Smart Region – Smart Community (grant number: CZ.02.1.01/0.0/0.0/17_048/0007435).

Disclosure statement

No potential conflict of interest was reported by the authors.

Funding

This work was supported by COST Action LAND4FLOOD (CA16209) supported by COST (European Cooperation in Science and Technology), www.cost.eu [grant number CA16209]; Operational Programme Research, Development and Education of the Czech Republic for financing the project Smart City – Smart Region – Smart Community [grant number CZ.02.1.01/0.0/0.0/17_048/0007435].

ORCID

Piotr Matczak http://orcid.org/0000-0002-8638-0141

References

Aerts, J. C. J. H., Botzen, W. J., Clarke, K. C., Cutter, S. L., Hall, J. W., Merz, B., … Kunreuther, H. (2018). Integrating human behaviour dynamics into flood disaster risk assessment. *Nature Climate Change*, *8*, 193–199.

Brundiers, K., & Hallie, C. E. (2018). Leveraging post-disaster windows of opportunities for change towards sustainability: A framework. *Sustainability*, *10*(5), 1390.

Czech Flood Protection Strategy. (2000). Strategy for protection against floods in the Czech Republic. Retrieved from http://eagri.cz/public/web/file/365719/Strategy_for_protection_against_floods.pdf

Czech Recovery Funding Law. (2002). Zákon č. 12/2002 Sb. o státní pomoci při obnově území postiženého živelní nebo jinou pohromou. Retrieved from https://www.zakonyprolidi.cz/cs/2002-12

Davies, J. (2016). Economic analysis of the costs of flooding. *Canadian Water Resources Journal / Revue Canadienne des Ressources Hydriques*, *41*(3/4), 204–219.

Declaration. (1991). Visegrad Declaration 1991. Retrieved from http://www.visegradgroup.eu/documents/visegrad-declarations/visegrad-declaration-110412

Drennan, L., McGowan, J., & Tiernan, A. (2016). Integrating recovery within a resilience framework: Empirical insights and policy implications from regional Australia. *Politics and Governance*, *4*(4), 74–86.

EC. (2007/60/EC). Directive on the assessment and management of flood risks. European Community.

George, A. L., & Bennett, A. (2005). *Case studies and theory development in the social sciences*. Boston, MA: MIT Press.

Henstra, D., Thistlethwaite, J., Brown, C., & Scott, D. (2019). Flood risk management and shared responsibility: Exploring Canadian public attitudes and expectations. *Journal of Flood Risk Management*, *12*(1), e12346.

Hruška, V., Czapiewski, K. L., & Kovács, Z. (2015). Post-agricultural rural space of the Visegrad countries: Economies, entrepreneurship and policies. *Studia Obszarów wiejskich / Rural Studies*, *39*, 129–144.

Hungarian Natural Disaster Act. (2011). CLXVIII. törvény a mezőgazdasági termelést érintő időjárási és más természeti kockázatok kezeléséről.

Istvan, F. (2015). The tools and limitations of integrated regional and rural development in Hungary. *Ter es Tarsadalom*, *29/1*, 132–148.

Ivanová, E., & Masárová, J. (2018). Performance evaluation of the Visegrad Group countries. *Economic Research-Ekonomska Istrazivanja*, *31*(1), 270–289.

Kučera, P. (2012). Povodňová daň přinesla do rozpočtu čtyři miliardy. Retrieved from https://zpravy.aktualne.cz/ekonomika/ceska-ekonomika/povodnova-dan-prinesla-do-rozpoctu-ctyri-miliardy/r~i:article:731470/

Lemkowska, M. (2019). Environmental Liability Directive call for development of financial instruments: The issue of compulsory insurance. *Ekonomia i Prawo*, *17*, 383.

Linnerooth-Bayer, J., & Amendola, A. (2003). Introduction to special issue on flood risks in Europe. *Risk Analysis*, *23*(3), 537–543.

Medd, W., Deeming, H., Walker, G., Whittle, R., Mort, M., Twigger-Ross, C., … Kashefi, E. (2015). The flood recovery gap: A real-time study of local recovery following the floods of June 2007 in Hull, North East England. *Journal of Flood Risk Management*, *8*(4), 315–328.

Meijerink, S., & Dicke, W. (2008). Shifts in the public–private divide in flood management. *International Journal of Water Resources Development*, *24*(4), 499–512.

Moatty, A., & Vinet, F. (2016). *Post-disaster recovery: The challenge of anticipation*. E3S Web of Conferences 7/17003. 3rd European Conference on Flood Risk Management (FLOODrisk 2016). Retrieved from https://doi.org/10.1051/e3sconf/20160717003

Mohd, R., Abu Bakar, N., Hassan, S., & Hussain, A. H. (2017). Sustainable post-disaster recovery plan for flood victims in Gua Musang and Kuala Krai, Kelantan. *Pertanika Journal of Social Sciences & Humanities*, *25*, 1–11.

Newig, J., Challies, E., Jager, N., & Kochskämper, E. (2014). What role for public participation in implementing the EU floods directive? A comparison with the Water Framework Directive, early evidence from Germany and a research agenda. *Environmental Policy and Governance, 24*(4), 275–288.

Penning-Rowsell, E. C., & Priest, S. (2015). Sharing the burden of increasing flood risk: Who pays for flood insurance and flood risk management in the United Kingdom. *Mitigation and Adaptation Strategies for Global Change, 20*, 991–1009.

Polish Act on Social Assistance. (2004). Dz. U. 2004 Nr 64 poz. 593.

Polish Crisis Management Act. (2007). Ustawa z dnia 26 kwietnia 2007 r. o zarządzaniu kryzysowym, Dz.U. 2007 nr 89 poz. 590.

Polish Water Act. (2017). Ustawa z dnia 20 lipca 2017 r. - Prawo wodne, Dz.U. 2017 poz. 1566.

Priest, J. S., Penning-Rowsell, E. C., & Suykens, C. (2016). Promoting adaptive flood risk management: The role and potential of flood recovery mechanisms. E3S Web of Conferences 7/17005. Retrieved from https://www.e3s-conferences.org/articles/e3sconf/pdf/2016/02/e3sconf_flood2016_17005.pdf

Raschky, P. A., Schwarze, R., Schwindt, M., & Zahn, F. (2013). Uncertainty of governmental relief and the crowding out of flood insurance. *Environmental and Resource Economics, 54*, 179–200.

Raška, P. (2015). Flood risk perception in Central-Eastern European members states of the EU: A review. *Natural Hazards, 79*(3), 2163–2179.

Raška, P., & Dostál, P. (2017). Evolution of disaster relief law under multiple transformations: Progressive learning or walking in a circle? *Environmental Science and Policy, 76*, 124–130.

Sandink, D., Kovacs, P., Oulahen, G., & Shrubsole, D. (2016). Public relief and insurance for residential flood losses in Canada: Current status and commentary. *Canadian Water Resources Journal / Revue Canadienne des Ressources Hydriques, 41*(3/4), 220–237.

Slavíková, L. (2018). Effects of government flood expenditures: The problem of crowding-out. *Journal of Flood Risk Management, 11*(1), 95–104.

Slavíková, L., Raška, P., & Kopáček, M. (2018). Mayors and "their" land: Revealing approaches to flood risk management in small municipalities. *Journal of Flood Risk Management, 12*(3), 1–10.

Slavíková, L., Vojáček, O., & Smejkal, T. (2017). Artificial shortage of surface water: How can water demand management mitigate the scarcity problem? *Water and Environment Journal, 31* (1), 12–19.

Solín, Ľ, Madajová, M., & Skubinčan, P. (2015). Mitigating flood consequences: Analysis of private flood insurance in Slovakia. *Journal of Flood Risk Management, 11*(S1), 173–185.

Suykens, C., Priest, S. J., van Doorn-Hoekveld, W. J., Thuillier, T., & van Rijswick, M. (2016). Dealing with flood damages: Will prevention, mitigation, and ex post compensation provide for a resilient triangle? *Ecology and Society, 21*(4), 1.

Tate, E., Strong, A., Kraus, T., & Xiong, H. (2016). Flood recovery and property acquisition in Cedar Rapids, Iowa. *Natural Hazards, 80*(3), 2055–2079.

Thaler, T., & Hartmann, T. (2016). Justice and flood risk management: Reflecting on different approaches to distribute and allocate flood risk management in Europe. *Natural Hazards, 83*(1), 129–147.

Thaler, T., & Priest, S. (2014). Partnership funding in flood risk management: New localism debate and policy in England. *Area, 46*(4), 418–425.

Thomalla, F., Lebel, L., Boyland, M., Marks, D., Kimkong, H., Tan, S. B., & Nugroho, A. (2018). Long-term recovery narratives following major disasters in Southeast Asia. *Regional Environmental Change, 18* (4), 1211–1222.

Vari, A., Linnerooth-Bayer, J., & Ferencz, Z. (2003). Stakeholder views on flood risk management in Hungary's Upper Tisza Basin. *Risk Analysis, 23*(3), 585–600.

Wandycz, P. (1992). *The price of freedom: A history of East Central Europe from the middle ages to the present*. New York, NY: Routledge, Chapman & Hall.

Weikard, H., Kis, A., & Ungvári, G. (2017). A simple compensation mechanism for flood protection services on farmland. *Land Use Policy, 65*, 128–134.

Young, O. R., Lambin, E. F., Alcock, F., Haberl, H., Karlsson, S. I., McConnell, W. J., ... Verburg, P. H. (2006). A portfolio approach to analyzing complex human-environment interactions: Institutions and land change. *Ecology and Society*, *11*(2), 31.

Ženka, J., Pavlík, A., & Slach, O. (2017). Resilience of metropolitan, urban and rural regions: A Central European perspective. *Geoscape*, *11*(1), 25–40.

Financial recovery schemes in Austria: how planned relocation is used as an answer to future flood events

Thomas Thaler and Sven Fuchs

ABSTRACT

Damages due to extreme hydro-metrological events request for additional efforts to enhance the implementation of property level flood risk adaptation (PLFRA) measures. Although a highly effective long-term measure, the planned relocation of individuals from areas at risk is rarely considered as an adaptive response. We evaluate how financial recovery schemes are actually linked to planned relocation option in two study sites in Austria. In both areas, more than 500 households were offered a voluntary planned relocation directly after extreme flood events. We conducted a semi-structured qualitative approach with 16 key respondents in order to identify how current financial recovery schemes are linked to ongoing and past relocation processes. The results show that there exists a missing link between disaster-aid payments and compensation for planned relocation. Participants gained from the programmes and used disaster-aid compensation schemes to increase the economic value of their houses. In addition, the financial schemes were not used to encourage further sustainability transition processes at local level. These data demonstrate the uneven distribution of payments with strong socio-economic implications for those who take part in planned relocation process. We recommend a better link between financial disaster-aid compensation and voluntary payout programmes, especially to reduce the uneven socio-economic distribution during the recovery phase.

1. Introduction

The devastation resulting from extreme hydro-meteorological events has raised the question of how flood risk management policy should be organised to respond to a warmer climate. In particular, it is forecasted that the societal impacts of climate change will amplify existing risks and create new risks to the human-environmental system (IPCC, 2012). Recent events, such as hurricane Harvey, Irma and Maria in the United States of America, Hurricane Matthew in Haiti and flooding worldwide, such as in Nigeria, South-East Asia, and the United Kingdom, have demonstrated that risk of flooding is on the rise and coastlines are being endangered. Notwithstanding significant efforts towards reduced exposure to flood risk, such as the construction of technical mitigation (e.g. levees, flood-proofing adjustments in residential and non-residential buildings and natural flood management strategies), damage caused by hydro-meteorological hazards remains significant. This is due to the socio-economic landscape of the past and dynamic land-use changes (Fuchs, Keiler, & Zischg, 2015, 2017; Paprotny, Morales-Nápoles, & Jonkman, 2018). In particular, the so-called 'expanding bull's-eye effect' shows the spatio-temporal dynamics for exposed buildings, infrastructure and even people in hazard-prone areas. The number of exposed buildings and people has heavily increased in many regions over the past few decades, which has presented one of the major challenges for policymakers as this increase has not been mirrored by any changes in flood prevention within hazard-prone areas (Ashley, Strader, Rosencrants, & Krmenec, 2014). This asymmetry became most notable in the socio-economic impacts experienced following Hurricane Harvey in Texas in 2017, where changes in land use resulted from the expansion of urban areas like many places across the world. The monetary value of assets and high-risk properties has been affected (Fuchs, Röthlisberger, Thaler, Zischg, & Keiler, 2017; Hutton, Tobin, & Montz, 2018). Hence the emergence of the Sendai Framework for Disaster Risk Reduction 2015–2030 (UNISDR, 2015) and further policy documents and academic discussions is an effort to urge and request for investment in structural and non-structural measures to enhance the preparedness of communities and the integrity of their assets (Klein et al., 2019). This has focused the attention of policy makers, citizens and other stakeholders on the question: which programme of measures for hazard mitigation and recovery planning is needed to be selected and implemented in order to enhance individual preparedness; and who is responsible to pay for it? Here, a central question is which social contract is needed to coordinate public administration and citizens to define individual roles and responsibilities (O'Brien, Hayward, & Berkes, 2009; Thaler et al., 2019). Most pertinently, the social contract must define the stage of action or inaction (or under-reaction) by the government in current flood risk management policy (Mortreux et al., 2018).

As stated in the literature, there is a missing link between financial contribution (from public and private sources) and strategy use to reduce vulnerability (Priest, Penning-Rowsell, & Suykens, 2016; Suykens, Priest, van Door-Hoekveld, Thuillier, & van Rijswick, 2016), or even to encourage a sustainability transition process to a low carbon society (Tàbara, Jäger, Mangalagiu, & Grasso, 2019). Sustainable transition processes can be understood as 'long-term, multi-dimensional, and fundamental transformation processes through which established socio-technical systems shift to more sustainable modes of production and consumption' (Markard, Raven, & Truffer, 2012, p. 956); such as decarbonisation of buildings. There is a great opportunity, not without challenges, to boost preventive action and implement property level flood risk adaptation measures and even sustainability transition, such as installing renewable energy production, in affected communities (Knobloch, 2006; Micangeli, Michelangeli, & Naso, 2013). This is especially the case where the government uses a planned relocation strategy to cope with large-scale flooding events. In most examples, these voluntary property acquisitions (or buy-out programmes) come into play after an event as a policy tool for recovery and hazard mitigation (Tate, Strong, Kraus, & Xiong, 2016).

Questions related to the use of financial compensation and planned relocation is often overlooked in the current literature. So far, no previous study investigates the link between planned relocation, financial compensation, improved individual preparedness and sustainability transition. Given these research gaps, the aim of this paper is to explore interlinkages between disaster-aid financial recovery schemes and planned relocation compensation schemes and to analyse the outcome and effects of the financial payments in terms of the improvement of the individual preparedness of homeowners and businesses and/or to encourage a sustainability transition towards decarbonisation in the selected study sites. This paper will endeavour to contribute to our knowledge around policy coordination and their influence on the social impacts of flood risk management strategies to communities. This has been tested by way of an investigative study on strategy for planned relocation sites in Austria, showing the missing links between financial recovery and planned relocation processes and how they contribute to uneven levels of individual well-being.

The paper is organised as follows: chapter 2 provides a literature review on planned relocation in natural hazards risk management. Chapter 3 outlines an overview of study sites as well as the methodology used. Chapter 4 investigates the financial recovery and planned relocation policy in the selected study sites and shows the interlinkages between the planned relocation and disaster-aid programmes. Chapter 5 reflects the ongoing financial recovery programmes and initiatives as well as key pitfalls and opportunities for policymakers.

2. Background

2.1. Policy coordination and social equity

The term policy coordination, coherence or integration has gained increased political and scientific importance since New Public Management has become a popular management concept for public administration. Policy coordination enhances public administration to act in a more 'efficient' way to increase their 'effectiveness', such as to avoid contradicted policy goals, instruments and strategies (Candel & Biesbroek, 2016; Peters, 2018). In particular, policy coordination should be able to solve complex problems. Consequently, policy coordination would reduce conflicts and encourage synergies, as different departments would find common aims and strategies how to reach them as well as define responsibilities, power, resources, functions and tasks (Candel & Biesbroek, 2016; Peters, 2018).

Analysing policy coordination can be distinguished between three main research questions. A first research question is focusing on motivations for policy coordination. Here, the main emphasis is on encouraging or discouraging policy coordination. The reality demonstrates that many policy strategies often act isolated and follow contradict goals, because departments fear of losing power and resources (people or budget), are specialist in one aspect, performance parameters does not allow cooperation, ideologies, or lack of political will and capacities. These factors often disintegrate the different departments within public administration (Peters, 2018). Consequently, these strong barriers in the policy coordination often cause unclear and undefined decisions about responsibilities. This cause additional resources to solve the problems of the past, such as reduction in exposure might request the use of planned relocation 'solve' this lack of coordination with all its negative consequences.

A second research question is focusing on the process of policy coordination, which can be distinguished between four main dimensions: (1) policy framework, (2) involvement and density of collaboration of different actors in the decision-making process, (3) policy goals and (4) policy instruments (Candel & Biesbroek, 2016). In flood risk management, 'optimal' policy coordination would avoid the use and introduction of perverse subsidies, goals and instruments, which increase exposed buildings in floodplain areas (Fuchs et al., 2017).

A third research question is focusing on performance (or effectiveness) of policy coordination. Here, a key question is if policy coordination might encourage the goal of social equity. At its most basic, the question of social justice concerns questions on the allocation of resources, capital and wealth across different members of a community or country. There are many different philosophical schools, which dictate different interpretations of just resource allocation and distributions (e.g. financial contribution, vulnerability, exposure, responsibilities), engagement in decision-making process (e.g. access to information, to court, legitimacy) and recognition (e.g. respect, discrimination). Usually, material unequal distribution reflects income and property ownership of each individual citizen. However, unequal economic distribution means barriers to access at labour market, education system, health care, lack of information, possibility to engage in decision-making process or get discriminate by public administration, etc. Social injustice such as living in floodplain areas is often based on socio-economic and socio-cultural inequality. In the discussion of justice and flood protection, not only the actual allocation of flood protection measures is significant, but also the way in which this allocation is achieved and how actors are recognise from politics (Thaler, Zischg, Keiler, & Fuchs, 2018; Thaler & Hartmann, 2016).

2.2. Financial flows in flood risk management

Currently, flood risk management heavily focuses on prevention (use of structural and non-structural measures) to reduce risk of future loss; for example, the designation of retention areas, flood-proofing of buildings and implementation of sustainable urban drainage measures (Holub, Suda, & Fuchs, 2012; Sayers et al., 2015; Thaler, Priest, & Fuchs, 2016). The consequence is that public administration usually prefers structural alleviation schemes to protect residential and non-residential properties (Thaler et al., 2016; Thaler & Hartmann, 2016). However, in many cases national governments are no longer able to satisfy engineering solutions to ensure the demand for safety. Key restrictions can arise from financial limitations due to financial and economic crises, availability of land, technical restrictions and limitations, or socio-demography instability whereby communities with strong out-migration become more financially vulnerable regarding the maintenance of flood alleviation schemes (Löschner et al., 2017; Thaler et al., 2016). A crucial factor affecting flood risk management is financial contribution (Fuchs, 2009), or distribution, which is in many cases organised through taxpayer funds (Thaler et al., 2018). Most importantly, costs for implementation and maintenance of flood alleviation schemes are often sourced from the general public budget and rarely from private–public partnerships (Thaler & Hartmann, 2016). One reason for this is the high investment costs of flood alleviation schemes (Thaler et al., 2018).

An exception, in many countries worldwide, is a financial scheme for recovery from natural hazard events. Residential and non-residential properties in many cases receive compensation (disaster relief) via public administration, private donor payments or flood hazards insurance (Booth & Tranter, 2018; Seifert-Dähnn, 2018). This extra assistance is often needed and is essential to compensate for and recover from the losses brought about by flood events. Many natural disasters have caused individuals and families to be overwhelmed, rendering them powerless and unable to respond adequately during the recovery period (Edgeley & Paveglio, 2017; Quarantelli, 1986). Consequently, recovery aid plays an important role in supporting these people. It should be noted that the ability of an individual or business to recover from a flood event is highly dependent on the financial means at the disposal of that individual or business. The speed of recovery is a reflection of the lapse time in receiving compensation; faster payments can include an earlier recovery (Twigger-Ross et al., 2014). Financial schemes for flood recovery do not tend to provide further regulations to encourage homeowners or businesses to engage in adaptive action. The contribution of financial recovery schemes towards a build-back-better model to enhance individual preparedness is often not provided for in the schemes (Boustan, Kahn, & Rhode, 2012; Priest et al., 2016 Raschky & Weck-Hannemann, 2007).

2.3. Planned relocation in flood risk management

In the past decades, various countries around the globe have used planned relocation as an option to reduce exposure as well as vulnerability (Usamah & Haynes, 2012). Examples of planned relocation can be seen in different countries across the Globe. However, these constitute exceptional cases in present-day flood risk management policy and practices. Nevertheless, one long-term strategy (the planned relocation of individuals, businesses and infrastructure) has been largely ignored as a possible option for inclusion in the national flood risk management policies across the world (Bukvic, 2015a, 2015b; Mortreux et al., 2018; Orlove, 2005; Perry & Lindell, 1997). These empirical cases demonstrate that planned relocation is a challenge process that goes beyond mere provision of alternative land and housing (Okada, Haynes, Bird, van den Honert, & King, 2014). These initiatives are usually conducted in the recovery period with large financial supports in place, which are defined by current governance frameworks.

Planned relocation is defined as

> a solutions-oriented measure, involving the State, in which a community (as distinct from an individual/household) is physically moved to another location and resettled there. Under this schematic approach, evacuation is distinct from planned relocation, and it does not fall within its scope. Planned relocation may, of course, play a role following evacuations in circumstances where places of origin are no longer habitable, and continued presence in the place of evacuation is not feasible. (UNHCR, 2014, p. 10)

Planned relocation of hazard-prone residents is a passive instrument in the context of flood risk management, aiming at a permanent reduction of potential damage (de Sherbinin et al., 2011). Usually, planned relocation is conducted after an event (Tate et al., 2016). It includes the physical process of displacing people to new living areas (Thaler, 2017). The aim is to preempt and correct past land-use developments with the goal of reducing potential for future flood damage. However, a lack of consideration is mainly based in the anticipation of low acceptance and public outcry among at-risk residents and decision-makers, who would be required in some cases to undergo radical change in their individual livelihoods and community development (Binder, Barile, Baker, & Kulp, 2019; López Carr & Marter Kenyon, 2015). Similar results can be found in many other fields, such as bioenergy or hydro-electricity projects or mining-induced relocation processes (Delang & Toro, 2011; Terminski, 2015).

Additionally, a key challenge is that planned relocation might also cause an increase in social and physical vulnerability of the community if new settlements do not take into account in their plans and designs preventative measures against future flood hazard events (Nalau & Handmer, 2018). As such, planned relocation is often based on short-term decisions, but can have long-term impacts for the community (Riad & Norris, 1996; Sipe & Vella, 2014). Main barriers are often: (a) lack of political will, (b) lack of regulations and financial compensation, (c) uncertainty about the process and outcome of the planned relocation and (d) psychological factors (transfer trauma), such as emotional reactivity, grief and exhaustion (Doherty & Clayton, 2011; Hino, Field, & Mach, 2017; Thaler, 2017).

Planned relocation impacted residents'travel time to work, and even forced some to change professions, especially in the Global South were people might move from formal to informal labor contracts or disregard informal settlers' needs within the planned relocation process (Ingram, Franco, Rumbaitis-del Rio, & Khazai, 2006; McCallin & Scherer, 2015). Nevertheless, positive examples can be found in various studies, such as McLeman and Smit (2006) or Bukvic and Owen (2017), where planned relocation show the highest effectiveness in reducing vulnerability within the society. Usamah and Haynes (2012) support this argument by the statement that planned relocation requires a strong collaboration with local citizens (in terms of planning and organising the relocation process) to fully consider their interests and motivation to move. People who are affected by planned relocation are confronted with profound changes in their lives (Perry & Lindell, 1997), such as disruption of existing social networks, emotional attachment to the house, financial constraints etc. In light of the primarily negative social costs and economic impacts expected by policymakers, there is a hesitation to propose relocation measures (Binder et al., 2019).

3. Methods and study site description

3.1. Study site description

Despite reservations mentioned in the literature review, the provincial government of Upper Austria has already conducted two planned relocation processes of private households, businesses and infrastructure along the River Danube, in the Machland basin (following the 2002 flood) and is currently administering large-scale relocations in the Eferding basin (following the 2013 flood). Both relocation initiatives were – and continue to be – accompanied by intensive, controversial stakeholder negotiations, public protest and media attention. The ongoing discourse on these two study sites has evoked multiple conflicts concerning the attribution of causes and responsibilities, and regarding the duties and actions of the actors involved (Schindelegger, 2018; Thaler, 2017).

The Machland basin experienced several severe flood events between 1991 and 2002. As a result, the public administration organised a first relocation process after the 1991 floods with minor response from the affected households. Directly after the large 2002 Danube floods, more than 254 households of the affected communities (Baumgartenberg,

Grein, Mitterbach, Saxen and St. Nikola) in the Austrian federal states Lower and Upper Austria were relocated with a total cost incurred of €92.5 million (Thaler, 2017). The second study site, Eferding basin, was heavily affected by the 2013 floods. In spring 2015, affected households were offered a compensation offer to move out of the hazard-prone area. Households had until the end of 2015 to decide whether or not they would accept this public compensation, but only 80 households agreed to relocate. As a consequence, the public administration renewed the offer and extended the deadline to December 2017. It is worthy of note that local stakeholders and citizens in Eferding basin have played an important role in the ongoing discourses and delays in the implementation process.

3.2. Methodology

In this study, we conducted a semi-qualitative structure of interviews with 16 key experts, at national, regional and local level, who were involved in local policy and administration between 2012 and 2017 (list of interviews and communication with stakeholders see Table 1 and Appendix 1). The aim of the interviews was to analyse their views on the relocation process, challenges, governance arrangements and policy discourses. In total, we performed 10 interviews[1] in the case of Machland basin between 2012 and 2014 and seven interviews in the case of Eferdinger basin between 2015 and 2017; some experts were interviewed for both study sites as they had played a crucial role in both areas. The interviews were conducted at two-year intervals as a follow-up wave to capture current barriers and drivers around the planned relocation implementation in both areas. The interview audio records were transcribed and analysed based on a grounded theory approach (Strauss & Corbin, 1998). The interviews focused on the governance arrangements, their role in the planned relocation process, legislative frameworks, financial compensation schemes and relief as well as the link to individual and business recovery. From this data, critical questions for using this adaptation policy arose. Additionally, we conducted a policy discourse analysis to assess published strategy and guidance documents to understand the formal and informal rules in disaster-aid compensation and relocation policy processes in Austria (full list of analysed policy documents see Appendix 2).

4. Results

4.1. Disaster-aid financial recovery schemes in Austria

The Austrian compensation policy provides for a mixture between public funding (*Katastrophenfonds*, Republik Österreich, 1996), private donor payments as well as in some cases recovery payments via private flood insurance (Fuchs, 2009; Rauter, Fuchs, Schindelegger,

Table 1. List of interviews.

Interview	Eferdinger Becken	Machland
National	2	4
Regional	2	1
Local	3	4
Total	7	9

Table 2. Public compensation framework in Austria.

Federal State	Min. level of compensation	Public compensation	Extension of compensation for low-income families	Insurance compensations
Burgenland	€ 2,000	20% – 30% (max. € 30,000)	No information	Insured losses are exempt from compensation
Carinthia	€ 440 (exceptions for low-income households)	No information about the level of compensation; compensation including private donor payments should not exceed 100% of the damage.	Yes, if the annual salary is under € 35,000 (net earnings)	Insured losses are exempt from compensation
Lower Austria	€ 1,000 (exceptions for low-income families)	20% (for flood events) – 50% (for debris flow events)	Yes, up to 70%	Insured losses are exempt from compensation
Salzburg	€ 1,000 (exceptions for low-income families)	30%	Yes, up to 60%; max. € 500,000	Insured losses are exempt from compensation
Styria	€ 1,000	Damage on buildings 50% (other damages 30%)	Yes	Insured losses are exempt from compensation
Tyrol	No information	20 – 50%	Yes, up to 80%	Insured losses are exempt from compensation
Upper Austria	€ 1,000	20% – 50%	Yes	Insured losses are exempt from compensation
Vorarlberg	€ 1,000 (insured buildings € 7,200)	50%	Yes, up to 75%	Insured losses are exempt from compensation
Vienna	Non-existent	Non-existent	Non-existent	Non-existent

& Thaler, 2019). In Austria, the insurance sector focuses mainly on private businesses (key focus on large business and not on small-medium business) and, only in some rare examples, covers damage to private buildings. Here, the insurance contract is part of household insurance (bundle system) where the level of compensation varies between €3,700 up to € 16,000 per contract; except for a small number of insurers who compensate up to 50% of the loss (Hanger et al., 2018; Holub & Fuchs, 2009). In addition, in many cases, only river floods or surface runoff are covered by insurance premiums and not groundwater flooding. On the other hand, businesses were offered more flexible contract conditions, where higher compensation against future losses is made available (Holub & Fuchs, 2009).

The Austrian government institutionalised the public compensation scheme for natural hazard losses, in contrast to other countries such as Germany, where ex-post-disaster compensation is arranged on an *ad hoc* basis (Hanger et al., 2018). The Austrian disaster fund is regulated under the Austrian Ministry of Finance (BMF) by legislation (Republik Österreich, 1996). Federal states are responsible for the implementation, matching the national payments and payout. However, there is no legal entitlement for citizens to directly access public compensation. As the Austrian flood risk management system is organised at federal state level, the federal states differ in the level and organisation of compensation (see Table 2). The level of compensation varied between 20% up to 50% of documented loss, and in some federal states the level of compensation rose to 80%. These discrepancies reflect social circumstances (such as low income, disability, single parent etc.) of the victims; similar to own retention of costs of natural hazards events. These points to the governmental aim to prevent social hardship and high social costs due to flood hazards. Ultimately, the key aim is to ensure the recovery of each household. Similarly, insurance compensation payments are excluded from the public compensation scheme, which might be another reason for the low market penetration (Hanger et al., 2018). Flood victims have no binding legal right to compensation, as federal states only implement compensation guidelines and not legal acts.

Besides the public compensation scheme, private donor payments play an important role in the Austrian recovery policy. Here, as stated in Table 2, some federal states have a number of rules surrounding the upper limit of compensation levels with regard to private donor payments. However, as interviewees stated, the key problem with private donor-aid payments is the uncertainty around timelines, levels of compensation and for what exactly the funds can be used (Interviewees 3, 4, 5). Interviewees stated that especially in the case of small events (with few houses affected), private contribution is not available, as these events do not receive broad media attention.

4.2. Policy framework for the planned relocation strategy in Austria

Austria has a long tradition in planned relocation programmes. Nevertheless, most relocation projects include the removal of a small number of households and businesses from hazard-prone areas. Exceptions include the planned relocation programmes along the River Danube in the past decades. Dating back to the 1970s, the Austrian government and the federal states (Lower and Upper Austria) used planned relocation as a tool to clear floodplain areas on an ongoing basis. In the past 20 years, the public government has focused on two main relocation projects in particular: in Upper Austria at Machland

and Eferdinger basin. The planned relocation framework included various aspects, such as levels of compensation, use of property rights, time of movement, payment rates etc. The relocation schemes have been voluntary insofar as the national and regional authorities offer compensation to the property at estimated market values. The level of compensation accounted for 80% (50% provided by the national government and 30% provided by the regional authorities) of the respective market value of a building for relocation. Furthermore, the government paid up to 80% of the demolition costs of the abandoned buildings. Another point to note is that householders and business who decided to move retain possession of the building plot, but are limited in their use of it. In all cases, the local authorities conducted a rezoning of the area from residential zones into grassland, with the main aim of restricting future construction in the areas to be abandoned, as there have been such cases in the past. However, the data highlighted in both study sites that – depending on the local authority – in most of the cases, the mayors commented on the challenges faced with rewilding the abandoned settlements (Interviews 6, 7, 8, 13, 14, 15, 16). Local authorities assigned dedicated building plots to the relocatees in order to avoid land speculations. However, the request of the relocatees to move to these designated building plots depends on the community; some accepted and some rejected the offer. The deadline for relocation was defined within a five-year period following the point of decision. Here, the interviewees stated that the goal was to avoid the creation of ghost towns in the area, as this had happened previously in the 1970s and 1980s relocation programmes in the federal state of Lower Austria. Lastly, the public administration provided two payment rates: 80% after the assigned contract and the remainder after the demolition of the building. Nevertheless, the main question remains: what happens with those people who refuse relocation and stay in the hazard-prone area? So far, the non-movers are restricted in the use of their buildings, such as no increase of actual and potential flood damage, which it remains unclear whether these householders are still eligible to opt into the financial recovery schemes of the public administration in the case of future flooding.

4.3. Role of financial compensation in the implementation scheme

First, as the interviewees stated there was no direct link between the financial recovery schemes and the strategy to implement property level flood risk adaptation (PLFRA) measures (Interviewees 1, 2, 3, 4, 5, 6). All compensation schemes (governmental compensation aid, private aid and insurance payments) do not encompass strategy use to reduce the vulnerability of residential or non-residential properties. In fact, there are no legislative rules or political guidance to change these administrative practices. Similar results can be found in terms of sustainability transition after the event. Principally, the financial recovery schemes do not provide for requests to increase energy efficiency or implement green roofs, as the compensation does not encompass any rules governing sustainability transition. In the same line is the voluntary compensation of property acquisition. The payments from the government did not provide for any requirement in terms of improving implement property level flood risk adaptation (PLFRA) measures or sustainability transition (such as decarbonisation) of new residential and non-residential buildings. As the interviewees stated; in most cases, there is little encouragement for new builds to be constructed in a sustainable way (Interviewees 13, 14, 15, 16). In both study sites, the empirical data from the policy documents revealed that there was no link between financial recovery scheme based on past events and property acquisition based on planned relocation. The outcome was that householders who invested in their homes after the event were provided with a higher compensation in the planned relocation process. The main reason was that the relocation payments were based on the market value of the building, which increased in recovery. Overall, the interviewees stated that especially low-income and elderly householders in the Eferdinger basin showed lower investment activity in the recovery period (Interviewees 6, 7, 8). The empirical results show the outcome of planned relocation policy encouraging an uneven distribution of payments with strong socio-economic implications for those who take part in the voluntary property acquisition. All interviewees stated the key missing link between financial recovery payments and planned relocation compensation was also based on a lack of political willingness (institutional barrier) to avoid this development and to encourage a more in-depth sustainability transition in the region. Additionally, the compensation payments included some further limitations. Firstly, there was a missing link with other natural hazards in the region; as relocatees (with a higher damage potential than previously) were victims of surface runoff events. This shows that relocation was not able to reduce the vulnerability of the communities. Analysing the data showed that in the Machland study site, the short distance relocation of the new buildings can be seen very critically as the new buildings can be found only outside the 1:100 floodplain area; in contrast to the Eferdinger basin example, where new buildings had to be built outside the 1:300 floodplain area (Interviewees 1, 2, 3, 5, 9, 10, 11, 12). This may mean further planned relocation strategies for the communities, given the residual risk based on extreme events and the impact of a warmer climate on future flood hazard events in Austria.

5. Discussion and conclusion

The primary objective of voluntary property acquisition is to reduce individual and societal risk in the region for future flood hazard events (Binder et al., 2019; Tate et al., 2016). This strategy is adopted most notably in areas where structural flood alleviation schemes cannot respond to the risks and challenges manifesting in the region; for technical, financial or political reasons. Therefore, planned relocation is often seen as a last resort (Hino et al., 2017). Recent extreme weather events showed that flood events in best-protected regions in fact exposed socio-economic vulnerabilities to as these areas, given the frequency with which they are faced with the challenge of residual risk. The residual risk of flood alleviation schemes cannot guarantee complete safety against extreme weather events (Thaler et al., 2016).

On the other hand, planned relocation has caused physical, mental and emotional stress as well as erasing social networks, the repercussions of which can still be felt today (Dannenberg, Frumkin, Hesss, & Ebi, 2019; Hino et al., 2017). The literature shows that relocatees are living in inferior conditions and have reduced livelihoods, when compared with the lives of those still living in the hazard areas (Riad & Norris, 1996). Additionally, planned relocation policy also has a strong impact on new living standards. For example, Thaler (2017) shows that mainly young farmers were able to increase their individual well-being, such as by creating new business activities or improving their quality of housing.

Against this background, the Austrian flood risk management policy implemented two large planned relocation schemes along the River Danube in the federal state of Upper Austria between 1991 and the present day. The goals of both relocation processes were (1) the reduction of risk and (2) the creation of further retention areas. Planned relocation was mainly used because of the limited use of structural flood alleviation schemes in the areas. The policy was implemented as a reactive policy, which is not unknown in the context of recovery (Ingram et al., 2006). The post-disaster planning was devised and implemented in an urgent and rapid way; without a long-term strategic planning process in terms of: (1) livelihood support, (2) regional development, (3) use/downgrade of infrastructure and (4) sustainability transition in the 'new' communities.

The first research objective of this paper focused on the interlinkages between disaster-aid financial recovery schemes and planned relocation compensation schemes. Our findings showed no linkages so far between these two financial payment schemes. A specific problem found was the missing coordination between the different sectoral policies acting in the recovery. In particular, the disaster-aid payments were focused on a fast recovery without a strategic view towards future flood risk management at local level. This is the result of a lack of policy coordination between water management (dealing with flood alleviation schemes) and (in most cases) agricultural authorities dealing with compensation payments. Consequently, the outcome in both study sites was an uneven use of policy provisions. In other words, families or businesses who used the financial resources provided by the disaster fund to reconstruct their property achieved higher compensation payments from the planned relocation process. Most importantly, the implication of this policy showed a strong distributional effect for homeowners and businesses in the study sites. An analysis of this policy, however, demonstrated that households with a high social vulnerability and a low individual capacity to respond to flood hazard events, and especially tenants with limited possibilities to reconstruct the property, were the main 'losers' affected by this missing link (Chakraborty, Collins, Montgomery, & Grineski, 2014). In fact, householders from repaired houses were more likely to take the compensation offer as the level of compensation was based on post-disaster market value of the building. The FEMA buyout programme of flooded properties usually foresees pre-disaster purchase prices. In addition, both programmes (Hazard Mitigation Grant Programme (HMGP) and the Community Development Block Grant (CDBG)) have the possibilities to add further incentives or bonuses. Additional, the public administration might acquire the ownership rights of the property in contrast to the Austrian planned relocation programme (Siders, 2019). Here, the U.S. programme allows a better response to individual needs and circumstances in contrast to the 'unify' planned relocation programme in Austria where income played no direct role in the level of compensation. In fact, high-income households with high-value houses got a higher level of compensation in compare to low-income families. Nevertheless, in practice also the U.S. programme compensation scheme shows that low-income communities often fail to achieve full financial compensation payments (Munoz & Tate, 2016).

The second research objective of this paper was to analyse the outcome and effects of the financial payments in terms of the improvement of the individual preparedness of homeowners and businesses and/or to encourage a sustainability transition towards decarbonisation of the area. The interviewees presented no radical change at the current level. These financial resources had no strategic plan to encourage individuals to prevent future losses or to change their individual behaviour, which supports the findings of Holub and Fuchs (2009), who have already criticised the missing link between preventive action and financial incentives in the Austrian flood risk management policy. We found that these missing opportunities (or lack of policy coordination) were also supported by a lack of strategic regional planning strategies in the study sites. So far, to our knowledge, no previous study has reported this missing link and its consequences.

In sum, the study showed two main policy implications. First of all, there is a need to build a better coordination and process-oriented view of flood recovery, in terms of linking financial disaster-aid payments with further flood risk management strategies to avoid social harm. Secondly, financial payments should address aspects of efficiency, in terms of improved individual preparedness as well as towards a sustainability transition within the ongoing build-back-better debate. In fact, the current policy encourage existing social inequality in flood risk management in Austria.

Note

1. I3 were interviewed for both study sites as he was involved in both planned relocation processes.

Acknowledgements

We would like to thank the two anonymous reviewers for their constructive comments.

Disclosure statement

No potential conflict of interest was reported by the authors.

Funding

This research received financial support from the Austrian Climate and Energy Fund and was carried out within the Austrian Climate Research Program (funding no. B567142).

ORCID

Thomas Thaler http://orcid.org/0000-0003-3869-3722
Sven Fuchs http://orcid.org/0000-0002-0644-2876

References

Ashley, W. S., Strader, S., Rosencrants, T., & Krmenec, A. J. (2014). Spatiotemporal changes in tornado hazard exposure: The case of the expanding bull's eye effect in Chicago, IL. *Weather, Climate, and Society*, *6*, 175–193. doi:10.1175/WCAS-D-13-00047.1

Binder, S. B., Barile, J. P., Baker, C. K., & Kulp, B. (2019). Home buyouts and household recovery: Neighborhood differences three years after Hurricane Sandy. *Environmental Hazards*, *18*, 127–145. doi:10.1080/17477891.2018.1511404

Booth, K., & Tranter, B. (2018). When disaster strikes: Under-insurance in Australian households. *Urban Studies*, *55*, 3135–3150. doi:10.1177/0042098017736257

Boustan, L. P., Kahn, M. E., & Rhode, P. W. (2012). Moving to higher ground: Migration response to natural disasters in the early twentieth century. *The American Economic Reviews*, *102*, 238–244.

Bukvic, A. (2015a). Integrated framework for the relocation potential assessment of coastal communities (RPACC): application to Hurricane Sandy affected areas. *Environment Systems and Decisions, 2015*, 264–278. doi:10.1007/s10669-015-9546-5

Bukvic, A. (2015b). Identifying gaps and inconsistencies in the use of relocation rhetoric: A prerequisite for sound relocation policy and planning. *Mitigation and Adaptation Strategies for Global Change, 2015*, 1203–1209. doi:10.1007/s11027-013-9532-5

Bukvic, A., & Owen, G. (2017). Attitudes towards relocation following Hurricane Sandy. Should we stay or should we go? *Disasters, 41*, 101–123. doi:10.1111/disa.12186

Candel, J. J. L., & Biesbroek, R. (2016). Toward a processual understanding of policy integration. *Policy Sciences, 49*(3), 211–231. doi:10.1007/s11077-016-9248-y

Chakraborty, J., Collins, T. W., Montgomery, M. C., & Grineski, S. E. (2014). Social and spatial inequities in exposure to flood risk Miami. *Florida. Natural Hazard Review, 15*, 04014006. doi:10.1061/(ASCE)NH.1527-6996.0000140

Dannenberg, A., Frumkin, H., Hesss, J. J., & Ebi, K. L. (2019). Managed retreat as a strategy for climate change adaptation in small communities: Public health implications. *Climate Change*, doi:10.1007/s10584-019-02382-0

Delang, C. O., & Toro, M. (2011). Hydropower-induced displacement and resettlement in the Lao PDR. *South East Asia Research, 19*, 567–594. doi:10.5367/sear.2011.0056

de Sherbinin, A., Castro, M., Gemenne, F., Cernea, M. M., Adamo, S., Fearnside, P. M., … Shi, G. (2011). Preparing for resettlement associated with climate change. *Science, 334*, 456–457. doi:10.1126/science.1208821

Doherty, T. J., & Clayton, S. (2011). The psychological impacts of global climate change. *American Psychologist, 66*(4), 265–276. doi:10.1037/a0023141

Edgeley, C. M., & Paveglio, T. B. (2017). Community recovery and assistance following large wildfires: The case of the Carlton complex fire. *International Journal of Disaster Risk Reduction, 25*, 137–146. doi:10.1016/j.ijdrr.2017.09.009

Fuchs, S. (2009). Susceptibility versus resilience to mountain hazards in Austria – Paradigms of vulnerability revisited. *Natural Hazards and Earth System Sciences, 9*, 337–352. doi:10.5194/nhess-9-337-2009

Fuchs, S., Keiler, M., & Zischg, A. (2015). A spatiotemporal multi-hazard exposure assessment based on property data. *Natural Hazards and Earth System Sciences, 15*, 2127–2142. doi:10.5194/nhess-15-2127-2015

Fuchs, S., Röthlisberger, V., Thaler, T., Zischg, A., & Keiler, M. (2017). Natural hazard management from a co-evolutionary perspective: Exposure and policy response in the European Alps. *Annals of the American Association of Geographers, 107*, 382–392. doi:10.1080/24694452.2016.1235494

Hanger, S., Linnerooth-Bayer, J., Surminski, S., Nenciu-Posner, C., Lorant, A., Ionescu, R., & Patt, A. (2018). Insurance, public assistance, and household flood risk reduction: A comparative study of Austria, England and Romania. *Risk Analysis, 38*, 680–693. doi:10.1111/risa.12881

Hino, M., Field, C. B., & Mach, K. J. (2017). Managed retreat as a response to natural hazard risk. *Nature Climate Change, 7*, 364–370. doi:10.1038/nclimate3252

Holub, M., & Fuchs, S. (2009). Mitigating mountain hazards in Austria – legislation, risk transfer, and awareness building. *Natural Hazards and Earth System Sciences, 9*, 523–537. doi:10.5194/nhess-9-523-2009

Holub, M., Suda, J., & Fuchs, S. (2012). Mountain hazards: Reducing vulnerability by adapted building design. *Environmental Earth Sciences, 66*(7), 1853–1870. doi:10.1007/s12665-011-1410-4

Hutton, N. S., Tobin, G. A., & Montz, B. E. (2018). The levee effect revisited: Processes and policies enabling development in Yuba County. *California. Journal of Flood Risk Management*, doi:10.1111/jfr3.12469

Ingram, J. C., Franco, G., Rumbaitis-del Rio, C., & Khazai, B. (2006). Post-disaster recovery dilemmas: Challenges in balancing short-term and long-term needs for vulnerability reduction. *Environmental Science & Policy, 9*, 607–613. doi:10.1016/j.envsci.2006.07.006

IPCC, Intergovernmental Panel on Climate Change. (2012). *Managing the risks of extreme events and disasters to advance climate change adaptation. A special report of working groups I and II of the intergovernmental panel on climate change*. Cambridge: Cambridge University Press.

Klein, J. A., Tucker, C. M., Steger, C. E., Nolin, A., Reid, R., Hopping, K. A., … Yager, K. (2019). An integrated community and ecosystem-based approach to disaster risk reduction in mountain systems. *Environmental Science & Policy, 94*, 143–152. doi:10.1016/j.envsci.2018.12.034

Knobloch, D. (2006). *Valmeyer, Illinois – Operation fresh start: Using sustainable technologies to recover from disaster*. Butte: National Center for Appropriate Technology.

López Carr, D., & Marter Kenyon, J. (2015). Human adaptation: Manage climate induced resettlement. *Nature, 517*(7534), 265–267. doi:10.1038/517265a

Löschner, L., Herrnegger, M., Apperl, B., Senoner, T., Seher, W., & Nachtnebel, H. P. (2017). Flood risk, climate change and settlement development: A micro-scale assessment of Austrian municipalities. *Regional Environmental Change, 17*, 311–322. doi:10.1007/s10113-016-1009-0

Markard, J., Raven, R., & Truffer, B. (2012). Sustainability transitions: An emerging field of research and its prospects. *Research Policy, 41*(6), 955–967. doi:10.1016/j.respol.2012.02.013

McCallin, B., & Scherer, I. (2015). *Urban informal settlers displaced by disasters: Challenges to housing responses*. Geneva: UNHCR.

McLeman, R., & Smit, B. (2006). Migration as an adaptation to climate change. *Climate Change, 76*, 31–53. doi:10.1007/s10584-005-9000-7

Micangeli, A., Michelangeli, E., & Naso, V. (2013). Sustainability after the thermal energy supply in emergency situations: The case study of Abruzzi Earthquake (Italy). *Sustainability, 5*, 3513–3525. doi:10.3390/su5083513

Mortreux, C., de Campos, R. S., Adger, W. N., Ghosh, T., Das, S., Adams, H., & Hazra, S. (2018). Political economy of planned relocation: A model of action and inaction in government responses. *Global Environmental Change, 50*, 123–132. doi:10.1016/j.gloenvcha.2018.03.008

Munoz, C. E., & Tate, E. (2016). Unequal recovery? Federal resource distribution after a Midwest flood disaster. *International Journal of Environmental Research and Public Health, 13*(5), 507. doi:10.3390/ijerph13050507

Nalau, J., & Handmer, J. (2018). Improving development outcomes and reducing disaster risk through planned community relocation. *Sustainability, 10*, 1–14. doi:10.3390/su10103545

O'Brien, K., Hayward, B., & Berkes, F. (2009). Rethinking social contracts: Building resilience in a changing climate. *Ecology & Society, 14*, 12.

Okada, T., Haynes, K., Bird, D., van den Honert, R., & King, D. (2014). Recovery and resettlement following the 2011 flash flooding in the Lockyer Valley. *International Journal of Disaster Risk Reduction, 2014*, 20–31. doi:10.1016/j.ijdrr.2014.01.001

Orlove, B. (2005). Human adaptation to climate change: A review of three historical cases and some general perspectives. *Environmental Science & Policy, 8*, 589–600. doi:10.1016/j.envsci.2005.06.009

Paprotny, D., Morales-Nápoles, O., & Jonkman, S. N. (2018). HANZE: A pan-European database of exposure to natural hazards and damaging historical floods since 1870. *Earth System Science Data, 10*, 565–581. doi:10.5194/essd-10-565-2018

Perry, R. W., & Lindell, M. K. (1997). Principles for managing community relocation as a hazard mitigation measure. *Journal of Contingencies and Crisis Management, 5*(1), 49–59. doi:10.1111/1468-5973.00036

Peters, B. G. (2018). The challenge of policy coordination. *Policy Design and Practice, 1*(1), 1–11. doi:10.1080/25741292.2018.1437946

Priest, J. S., Penning-Rowsell, E. C., & Suykens, C. (2016). Promoting adaptive flood risk management: the role and potential of flood recovery mechanisms. E3S Web of Conferences 7/17005.

Quarantelli, E. (1986). What is disaster? The need for clarification in definition, conceptualization in research. In B. Sowder, & M. Lystad (Eds.), *Disasters and mental health: Selected contemporary perspectives* (pp. 49–81). Rockville, MD: National Institute of Mental Health.

Raschky, P. A., & Weck-Hannemann, H. (2007). Charity hazard – A real hazard to natural disaster insurance? *Environmental Hazards, 7*, 321–329. doi:10.1016/j.envhaz.2007.09.002

Rauter, M., Fuchs, S., Schindelegger, A., & Thaler, T. (2019). Breaking down the legal framework for flood protection in Austria: Individual and state responsibilities from a planning perspective. *Water International, Online First*, doi:10.1080/02508060.2019.1627641

Republik Österreich. (1996). *Bundesgesetz über Maßnahmen zur Vorbeugung und Beseitigung von Katastrophenschäden (Katastrophenfondsgesetz 1996 – KatFG 1996). BGBl 201/1996*. Vienna: Republik Österreich.

Riad, J. K., & Norris, F. H. (1996). The influence of relocation on the environmental, social, and psychological stress experienced by disaster victims. *Environment and Behavior, 28,* 163–182. doi:10.1177/0013916596282001

Sayers, P. B., Galloway, G., Penning-Rowsell, E. C., Shen, F., Wen, K., Chen, Y., & Le Quesne, T. (2015). Strategic flood management: Ten 'golden rules' to guide a sound approach. *International Journal of River Basin Management, 13,* 137–151. doi:10.1080/15715124.2014.902378

Schindelegger, A. (2018). Relocation for flood retention in Austria. In E. Hepperle, J. Paulsson, V. Maliene, R. Mansberger, A. Lisec, & S. Guelton (Eds.), *Opportunities and constraints of land management in local and regional development* (pp. 111–120). Zürich: vdf Hochschulverlag.

Seifert-Dähnn, I. (2018). Insurance engagement in flood risk reduction – examples from household and business insurance in developed countries. *Natural Hazards and Earth System Sciences, 18,* 2409–2429. doi:10.5194/nhess-18-2409-2018

Siders, A. R. (2019). Social justice implications of US managed retreat buyout programs. *Climatic Change, 152*(2), 239–257. doi:10.1007/s10584-018-2272-5

Sipe, N., & Vella, K. (2014). Relocating a flood affected community: Good planning or good politics? *Journal of the American Planning Association, 80,* 400–412. doi:10.1080/01944363.2014.976586

Strauss, A., & Corbin, J. M. (1998). *Basics of qualitative research: Techniques and procedures for developing grounded theory.* Thousand Oaks: SAGE.

Suykens, C., Priest, S. J., van Door-Hoekveld, W. J., Thuillier, T., & van Rijswick, M. (2016). Dealing with flood damages: Will prevention, mitigation and ex post compensation provide for a resilient triangle? *Ecology & Society, 21,* 1. doi:10.5751/ES-08592-210401

Tàbara, J. D., Jäger, J., Mangalagiu, D., & Grasso, M. (2019). Defining transformative climate science to address high-end climate change. *Regional Environmental Change, 19,* 807–818. doi:10.1007/s10113-018-1288-8

Tate, E., Strong, A., Kraus, T., & Xiong, H. (2016). Flood recovery and property acquisition in Cedar Rapids, Iowa. *Natural Hazards, 80,* 2055–2079. doi:10.1007/s11069-015-2060-8

Terminski, B. (2015). *Development-induced displacement and resettlement: Causes, consequences, and socio-legal context.* IDIDEM: Stuttgart.

Thaler, T. (2017). The challenges with voluntary resettlement processes as a need under changing climate conditions. In F. van Straalen, T. Hartmann, & J. Sheehan (Eds.), *Property rights and climate change. Land use under changing environ-mental conditions* (pp. 25–37). Abington: Routledge.

Thaler, T., Attems, M.-S., Bonnefond, M., Clarke, D., Gatien-Tournat, A., Gralepois, M., … Fuchs, S. (2019). Drivers and barriers of adaptation initiatives – How societal transformation affects natural hazard management and risk mitigation in Europe. *Science of the Total Environment, 650,* 1073–1082. doi:10.1016/j.scitotenv.2018.08.306

Thaler, T., & Hartmann, T. (2016). Justice and flood risk management: Reflecting on different approaches to distribute and allocate flood risk management in Europe. *Natural Hazards, 83,* 129–147. doi:10.1007/s11069-016-2305-1

Thaler, T., Priest, S., & Fuchs, S. (2016). Evolving interregional co-operation in flood risk management: Distances and types of partnership approaches in Austria. *Regional Environmental Change, 16,* 841853. doi:10.1007/s10113-015-0796-z

Thaler, T., Zischg, A., Keiler, M., & Fuchs, S. (2018). Allocation of risk and benefits—distributional justices in mountain hazard management. *Regional Environmental Change, 18,* 353–365. doi:10.1007/s10113-017-1229-y

Twigger-Ross, C., Kashefi, E., Weldon, S., Brooks, K., Deeming, H., Forrest, S., … Tapsell, S. M. (2014). *Flood resilience community pathfinder evaluation: Rapid evidence assessment.* London: Defra.

UNHCR. (2014). *Planned relocation, disasters and climate change: Consolidating good practices and preparing for the future.* Geneva: UNHCR.

UNISDR, United Nations Office for Disaster Risk Reduction. (2015). *Sendai framework for disaster risk reduction 2015–2030.* Sendai: UNISDR.

Usamah, M., & Haynes, K. (2012). An examination of the resettlement program at Mayon Volcano: What can we learn for sustainable volcanic risk reduction? *Bulletin of Volcanology, 74*(4), 839–859. doi:10.1007/s00445-011-0567-8

Appendices

Appendix 1: complete list of interviews and communications with stakeholders

Interview 1: National Authority on Eferdinger basin.
Interview 2: National Authority on Eferdinger basin.
Interview 3: Regional Authority on Eferdinger and Machland basin.
Interview 4: Regional Authority on Eferdinger basin.
Interview 5: Regional Authority on Eferdinger basin.
Interview 6: Local authority on Eferdinger basin.
Interview 7: Local authority on Eferdinger basin.
Interview 8: Local authority on Eferdinger basin.
Interview 9: National Authority on Machland basin.
Interview 10: National Authority on Machland basin.
Interview 11: National Authority on Machland basin.
Interview 12: National Authority on Machland basin.
Interview 13: Local Authority on Machland basin.
Interview 14: Local Authority on Machland basin.
Interview 15: Local Authority on Machland basin.
Interview 16: Local Authority on Machland basin.

Appendix 2: complete list of used policy documents

Used policy documents (April 2019)

Federal States disaster-aid documents

Burgenland Allgemeine Richtlinien zur Förderung der Behebung von Katastrophenschäden Konsolidierte Fassung gem. Beschluss der Burgenländischen Landesregierung vom 27.06.2017.

Carinthia Richtlinien für die Durchführung von Hilfsmaßnahmen des Kärntner Nothilfswerkes ab 27.06.2017.

Lower Austria Beihilfen für Gemeinden zur Behebung von Katastrophenschäden.

Upper Austria Förderung der Behebung von Katastrophenschäden im Vermögen physischer und juristischer Personen mit Ausnahme die der Unternehmen und die der Gebietskörperschaften LFW-2016-288692/6.

Salzburg Richtlinien für die Gewährung einer finanziellen Beihilfe des Landes zur Behebung von Katastrophenschäden im Vermögen natürlicher und juristischer Personen mit Ausnahme von Gebietskörperschaften.

Styria Richtlinie für die Abwicklung des Entschädigungsverfahrens nach Katastrophenschäden im Vermögen natürlicher und juristischer Personen mit Ausnahme der Gebietskörperschaften im Bundesland Steiermark – Katastrophenfonds-Richtlinie Steiermark.

Tyrol Leitlinien für die Beantragung von Beihilfen aus dem Katastrophenfonds.

Vorarlberg Richtlinie der Vorarlberger Landesregierung für die Gewährung von Beihilfen zur Behebung von Elementarschäden.

National Audit Office (2017): Bericht des Rechnungshofes Katastrophenhilfe in Niederösterreich, Salzburg und Tirol III–53 der Beilagen zu den Stenographischen Protokollen des Nationalrates XXVI. GP.

The French *Cat' Nat'* system: post-flood recovery and resilience issues

Bernard Barraqué and Annabelle Moatty

ABSTRACT

Successive French governments have progressively decentralised flood control policy to increase the role of local authorities in planning and crisis management. In 1982, a law mandated local risk maps for 5 types of exceptional natural hazards and set up Cat' Nat', a national system of damage compensation based on an insurance super-fund. While this system clearly improved the situation of victims of extreme events through subsidies in housing and infrastructure reconstruction, it did not necessarily foster a parallel reduction of vulnerability: insurance is more tuned to the past than to the future, and the tacit rule supports identical reconstruction so as not to increase the pre-disaster vulnerability (but not reducing it either). Yet indirectly, the recognition of a state of natural disaster triggers vulnerability reduction later, through various measures at various scales, from housing level (*build back better*) to the PAPI (action programmes for flood prevention); we describe them and present a case study before presenting hypotheses to explain the resistance of private landowners as well as potential improvements to better bridge recovery and resilience.

Introduction

In France, flood control (the most important natural hazard both in terms of the number of events and the scale of economic impact) has been framed by specific central-local relationships, like most territorial policies. The Constitution of the Vth Republic adopted in 1958 increased the centralised nature of the regime, but met with resistance from local government, giving rise to what Michel Crozier and his disciples, Pierre Grémion (1976), Jean-Pierre Worms (1966) called 'cross regulation': central government developed a policy to modernise the economy, but at local level, society would resist the proposed change, unless the government would bring financial support. Without this support it would have to postpone the reforms or give dispensations. Bargaining took place between the 'prefect' (*préfet*, head of central administration at *département* – county – level), and the county's political leaders and aldermen. International comparisons fostered by the rise of European water policy then triggered decentralisation as a way to make territorial authorities more responsible.

The decentralisation policy initiated in 1982 included an important reform of the control of 5 natural hazards: floods, landslides, avalanches, earthquakes and volcano eruptions (the latter in overseas counties). It created a national funding mechanism generated from additional insurance premiums paid by all households and vehicle owners, to be mobilised to cover damages resulting from natural disasters (exceptional events). As a counterpart, local authorities in risk areas were required to draft 'risk exposure maps' to ban construction or subject it to vulnerability reduction measures in the disaster prone parts of their territory. Repeated severe droughts more recently led to add a sixth eligible disaster: housing damages due to clay soil subsidence.

According to Lamond and Proverbs (2009), the notion of resilience encompasses pre-disaster planning and warning systems, emergency handling procedures and post-disaster reconstruction. In this article we focus on the link between recovery and resilience; we first present the context of flood control and land-use policies in the country; then we discuss the way post-disaster recovery procedures operate, and how they improve the situation for victims of natural disasters; and follow this by questioning whether the recovery system also triggers resilience measures, in particular we ask 'at household level, is recovery leading to "building back better"?' Finally, we illustrate the presentation with a field case study, before concluding with potential improvements to the policy.

In other words, we address four questions: how far was decentralisation of flood control driven and why was it incomplete? Is post-disaster recovery now improved and how much does it cost? Does the insurance-based recovery funding improve resilience? And how can landowners be better involved in vulnerability reduction?

Decentralisation of water policy and flood control in France

As early as the 1960s, a regionalisation policy initiated a new form of governance, made more transparent and open to civil society through public participation mechanisms (Duran & Thoenig, 1996). In the water policy sector, this regionalisation included the creation of 6 *Agences de l'eau* and 6 *Comités de bassin* (institutions at the river basin district level) covering the country, and operating under a mutualised version of the polluter-pays principle: i.e. that water users degrading the quality or the quantity of water in the environment pay a collectively agreed upon levy to the institution, which uses the money to subsidise environmentally friendly projects proposed by pro-active water users. Over the following decades, these river-basin institutions helped the learning process about valuing water and turning water policy into one focussed on usership rather than on ownership, *i.e.* separating the right to use water from landownership rights, an important change in the country which invented the Civil Code. Interestingly, French river-basin institutions were not entrusted with flood control measures: they remained a prerogative of central government (Barraqué, 2014) despite the overlap between the recovery of the aquatic environment and flood control, as some land-based measures support both.

Decentralisation was promoted further with laws voted in 1982–83, under the left wing government. New planning laws gave competence for granting building permits to local authorities, on the condition that they first set up a land-use plan, for review by the government's services at county level. The idea was to decentralise urban planning at local level, but to limit the possible subjections of elected councils to vested interests and land-use based speculation, as well as to check the incorporation of national regulations. But rarely would a prefect take a local authority to court for illegal planning decisions.

In the same spirit in 1982, a law was passed to launch a new recovery and vulnerability reduction policy for the above-mentioned 5 types of hazards.[1] To fund the recovery, rather than use the budget of the *Agences de l'eau*, the law mobilised insurance companies to provide compensation to victims from a special insurance fund called Cat' Nat' (abbreviation for *Catastrophe Naturelle*) (Figure 1), set up thanks to an addition on all insurance premiums at national level; for vulnerability control, the law mandated all appropriate local authorities to draft risk maps, the *'Plans de prevention du risque'* (risk prevention plans, PPR, and for floods, PPR-*inondation* or PPRi); however, the plans were not done well or quickly; in addition long term precaution and environmental issues could not compete with short term added value of urban development. In 1987, through additional legislation, the prefects recovered the responsibility to draw the PPR, but difficulties remained in their completion and they were not rapidly incorporated into local town plans.

Central government services eventually found themselves with flood events responsibilities, as illustrated by the Xynthia disaster (47 casualties in total): in February 2010 on the Atlantic coast in La Faute-sur-Mer, close to La Rochelle, a severe depression provoked a sea surge which broke the dykes and submerged housing estates on floodable land (Vinet, Boissier, & Defossez, 2011). Families of the victims brought a lawsuit against the municipality of La Faute-sur-Mer and central government. The municipality was condemned, criminally for the mayor; but central government was also condemned for its inaction against a town plan which favoured development in the hazard prone areas, and for insufficient maintenance of the dykes.

Today, the implementation of the Floods Directive[2] (FD – 2007/60/EC) is the responsibility of the prefects and local authorities, as in the past, i.e. at the level of administrative territories. However, in the years before the FD's adoption in 2007, catchment-based institutions called *Etablissements Publics Territoriaux de Bassin* (Public Territorial Catchment Institutions – EPTB) were created at a smaller scale than the 6 river basin districts, and some of them developed a tool called *Programmes d'Action pour la Prevention des Inondations* (Flood Prevention Action Programs – PAPI), to financially support projects aimed at reducing vulnerability and 'returning space' to the rivers; which has become an essential part of flood risk management policy, as illustrated below.

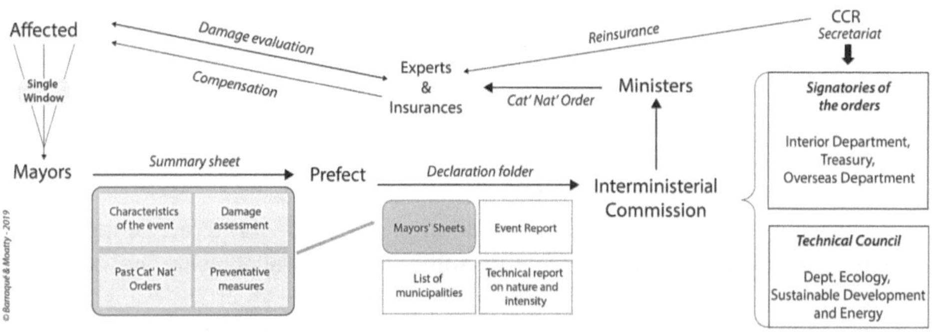

Figure 1. The Cat' Nat' damage compensation system (source: authors' elaboration).

More recently in 2014, and possibly following the Xynthia lawsuit, the government decided to transfer the competency for aquatic environment management and flood protection at local level, by organising a grouping of tiny local authorities into larger and more sustainable units called *Etablissements Publics de Cooperation Intercommunale* (Intercommunal Cooperation Public Institutions – EPCI). Again local authorities complained that these new tasks would be an unfunded mandate. The government allowed them to tax local residents and landowners up to 40 €/cap/yr to cover this new expense. But on 1 January 2019, only one third of EPCI had adopted this new tax (AdCF, 2019), with limited rates, since they are already pressed by the Cour des Comptes (national accounting office) to reduce budget spending in general …

The progressive decentralisation of the control of risks as presented here should be challenged with key questions related to recovery and resilience: does this evolution improve post-disaster recovery, and in addition does it foster vulnerability reduction? Before discussing this, a presentation of the recovery system and its effectiveness is needed.

Recovery with Cat' Nat': how does it work?

As mentioned above, Cat' Nat' works thanks to an additional fee on all insurance premiums in the country, whether policy holders are at risk or not. The resulting fund covers only the damages of extreme events, which were not covered before the 1982 law: insurance companies would then argue a case of *force majeure* (*i.e.* beyond what is statistically foreseeable, sometimes called 'an act of God' in the English speaking world).

In their typology of flood coverage systems at international level, Lamond and Penning-Rowsell have picked Cat' Nat' as the typical example of 'bundled insurance backed by the State' (Lamond & Penning Rowsell, 2014). The Cat' Nat' fund is created from 12% extra on the premiums paid to cover homes, businesses and some agricultural assets, and 6% on automobile premiums. 53.1% of the total comes from domestic residences, 37.3% from business risks, 3.4% from agriculture risks, and 6.2% from automobiles *i.e.* €100 million. If an insurance company has difficulties in facing its liabilities, it can get the support of the Caisse Centrale de Réassurance (Central Reinsurance Agency – CCR).

When an extreme event occurs, mayors of municipalities hit by a natural disaster can make a Cat' Nat' claim to the prefect (Figure 1); the prefect in turn reports to an inter-ministerial commission, responsible for deciding on the 'naturalness' and 'exceptional intensity' of the phenomenon, and whether 'the usual measures to avoid such damage could not prevent them to happen or could not be taken'.[3] Upon the commission's advice, the relevant ministers will eventually sign a Cat' Nat' order, acknowledging a state of natural disaster. Only then, are insurance companies liable to mobilise the additional premiums they have set aside through the super-fund, and cover the damages.

Post-disaster coverage includes the replacement of damaged infrastructure and buildings, as well as the complete restoration of services and economic revitalisation. This support through the insurance system can be supplemented with direct subsidies from the State: The government supports local authorities in dealing with the distress of victims on top of the insurance system, including compensation for uninsured local public assets. For instance, the *Fonds de Solidarité en faveur des Collectivités Territoriales* (Solidarity Fund in Support of Local Authorities – FSCT)[4] can be mobilised if the amount of damage to the local authority is between €150,000 and 6 million; beyond this level the *Fonds de Calamités Publiques* (Public Calamity Fund) can be called in. These funding measures, which are separate from Cat' Nat', play a decisive role, since they better encourage the adoption of preventive adaptation measures within reconstruction (Moatty, Gaillard, & Vinet, 2017).

The Cat' Nat' system prevents many insured families and businesses from being ruined, and in theory allows them to obtain quickly (within 3 months) the funds they need to repair their belongings, at least for those who are insured. In addition, after the event, insurance companies are in principle not allowed to raise the premiums of concerned insurees, given that it is exceptional. This can be compared with the situation in England where insurers may change the contract conditions for those flooded (Lamond, Proverbs, & Hammonds, 2009). Although based on a small set of examples, the following Tables 1 and 2 illustrates the improved damage coverage allowed by Cat' Nat' compared with 2016 data collected by Munich Re:

In 2017, the last year with available consolidated statistics, the total amount of additional premiums raised by insurance companies reached 1.641 bn €, *vs* 1.601 bn € in 2016. This budget is being used chiefly to cover two exceptional events: the first, cyclone Irma which hit the Caribbean islands of St Martin and St Barthélemy, and cost around 2 billion € of insured losses according to CCR; the second, the 2018 drought following the one of 2016 will certainly cost even more to the system.

Altogether in its 35 years' existence, the Cat' Nat' system has managed to finance the physical reconstruction of damaged assets at national level whilst increasing additional premiums paid to insurance companies by only 10%. Confidence in the system relies on public reinsurance system which supports insurance companies facing hardship, and collects the data allowing a good follow-up (CCR, 2018). On top of it, the whole system is guaranteed by central government budget which would typically be mobilised in case of a more devastating event like an earthquake hitting the Nice area.

Now, a potential accumulation of extreme events, related to climate change, questions the sustainability of the recovery system. The fee on insurance premiums has already been

Table 1. Cost and coverage rate of some 2016 disasters (Munich Re; French CCR).

Area	Event	Cost estimate	Coverage rate
World	Total natural disasters	€ 167 bn	30%
Japan	Earthquake April 14–16	€ 29.6 bn	19%
Europe	May-June storms	€ 5.7 bn	47%
France	May floods*	€ 1.4 bn	64%

*this was the second costliest event since the inception of Cat' Nat', with 150,000 victims and 5 casualties.

Table 2. Cat' Nat' key 2017 data (from CCR, 2018).

	2017	From 1982 (2000 for vehic.)	Yearly average
Number of insurance contracts (– including vehicles)	90 million		
Cat' Nat' income from additional premiums	1.64 Md€		
Cost of Cat' Nat' events on non-vehicle insured damages	IRMA 1.97 Md€ Drought 800 M €	33 Md€ (56% floods, 33% droughts)	936 M€
Cost of insured -vehicle damages		707 M€	40 M€

growing steadily since its inception: initially, the additional rate on non-vehicle premiums was set at 5.5%, but it was raised to 9% in 1985, to restore an interannual balance between income and expenses; and it rose again to 12% in 2000, because more severe and repeated droughts resulted in soil subsidence provoking damage to overlying properties. Today with the possibility that climate change increases the number and severity of extreme events, the system could potentially become bankrupt. In 2016, the total disbursement of insurance companies to victims of floods and droughts already exceeded the total budget gathered from insurees by 4.5%. For 2017, the deficit is even worse. It is still difficult to assess because the rate of insurance coverage is much lower in the French West Indies than on the mainland: the rate of housing insurance penetration is only 52% in the *départements d'outre-mer* (overseas counties – DOM) vs 99% in mainland France; but the cost of Irma alone represents more than 6% of total disasters' damage cost since the creation of Cat' Nat' in 1982. In 2018 an exceptional drought on most of France (and Northern Europe), together with a dramatic flash flood in Aude county in October will also probably leave the fund in deficit. It remains to be seen if implementation of the FD will lead to reviewing the PPRi and increasing land-use limitations, or if Cat' Nat' will need a new increase of the fee on insurance premiums.

From recovery to resilience: how and when

> Economic resilience depends on the capacity of a government to fund recovery and reconstruction through a large span of public and private mechanisms, like budget reallocations, tax increase, reserves mobilisation, national and international bonds, international (European) aids, insurance and re-insurance indemnities and financial obligations like 'catastrophe obligations'. (Mechler, Linnerooth-Bayer, Hochrainer, & Pflug, 2006; UNISDR, 2013)

The reconstruction period offers a 'window of opportunity' to reduce vulnerability (Christoplos, 2006; Moatty, 2017) This can take two different shapes, which can *a priori* appear contradictory, and yet in fact follow each other and set resilience as a dynamic process. Indeed, during the crisis, resilience means resisting, at household or territorial level; while in the longer post-disaster run, it means a capacity to bounce back to normal or to non-degraded functioning, eventually with adaptation (Moatty, Vinet, Defossez, Cherel, & Grelot, 2018).

But in France, the Cat' Nat' relative success has a negative counterpart: insurance companies unintendedly reduce resilience since they reimburse victims on the basis of actual damages incurred, which does not foster different and more resilient reconstruction: additional costs of vulnerability reduction are not well covered, as an official inquiry on post-disaster victims' compensation puts it: 'The risk prevention policy and the compensation of natural disasters are juxtaposed but they largely ignore each other' (Moatty & Vinet, 2016). On top of this, many damages are not considered by insurance companies: those on the public domain, outdoor housing damages, non-monetisable values (casualties, health impacts, emotional loss), etc. (Moatty et al., 2018). Vulnerability reduction has to be funded differently.

In the French system, resilience then relies more on prevention than on crisis management and post disaster measures, and prevention is chiefly introduced through the risk maps: the PPR. The PPR act as a counterpart of national reconstruction financing by making a local obligation to zone risk areas. Typically PPRi are maps showing areas exposed to a reference flood hazard (it is often the 100-year flood that is used, or the highest known water levels when they are higher), and include bans on building (in areas where the risk is considered strong), or restrictions on building (*e.g.* no valuables on ground floor) in other areas. This remains a touchy issue due to the political weight of private property, and PPR meet land-owners resistance.

And, since Cat' Nat' funding is national but PPR are local, obviously local authorities are tempted to underestimate the risks and to support economic and urban development at the expense of prevention, all the more so when they expect that losses will be covered by Cat' Nat'. Zoning regulations are feared by local representatives as impacting negatively on land values and the attractiveness of the commune, particularly where other natural or industrial risks are identified. According to business daily *Les Echos*,

> this very protective regime also has a perverse effect: it delays setting up efficient prevention policies, and even in some cases, it relieves actors from their responsibilities. Too many mayors allow house building in flood prone areas; too few coastal communes adopt plans on sea surge risks; and too few in general impose a geological survey before granting a building permit. (Maujean, 2018 – our translation)

Various experts question a perverse effect: 'automatic reimbursement' of disasters tends to reduce the victims' responsibility (Bourrelier, 1997; Ledoux, 2000; Lefrou et al., 2000).

This is also why the drafting of the plans was transferred from municipalities to prefects' services at county level in 1987, but implementation difficulties remain frequent, due to the 'crossed regulation' politics mentioned at the beginning of this paper.

Additional measures to reduce vulnerability

Floods however do trigger vulnerability reduction efforts, even though not necessarily in the recovery phase. Despite local resistance, the above-mentioned PPRi progressively cover all concerned communes. At the end of 2017, the number of approved PPR and other risk plans resulting from previous legislation exceeded respectively 20,000 (all risks) and 14,000 (floods and mudslides), covering 10,400 communes for flood risks. This means that in most communes at risk of flooding, there are some limitations on building in areas at risk. In addition, a 2004 law mandated municipalities to set up, within 2 years after the PPRi is approved, what is called a *Plan Communal de Sauvegarde* (Communal Safeguard Plan – PCS). This PCS establishes alert and crisis management procedures[5] such as evacuation and emergency resettlement. The law also mandates local authorities to issue a *document d'information communal sur les risques majeurs* (Communal Information Document on Major Risks – DICRIM), to help the local population be prepared if a disaster is announced. At the end of 2017, respectively 8000 and 6000 communes had set up their PCS and their DICRIM, which means that local knowledge about the level of risk has probably improved, even though in a variable manner: realising a PCS implies that the local authority either has staff qualified in environment and urban planning, or outsources to consultants, who make the vulnerability diagnosis and draft the immediate action sheets needed. In the latter case there is a chance that PCS remain unknown or non-appropriated by elected representatives, reflecting lack of interest or excessive standardisation due to economies of time and money.

Altogether, however, the post-disaster period is a time of increased consciousness of vulnerabilities, and thus of better acceptance of prevention's additional costs (Quarantelli, 1999). Improved knowledge of hazards and vulnerabilities lead central government and the insurance system together to update and reinforce regulations. To give an example, a review on *Build Back Better* incentives (MRN, 2018; MRN & et al., 2017) was made after the costly flood of May-June 2016 south of Paris. It concluded that the pre-existing PPRi should be reinforced: they had been designed after the worst event ever recorded, the flood of 1910, and yet in 2016 some cities recorded higher levels still (+40 cm in Nemours, +30 cm in Montargis). Detailed observation at building level showed that 95% of the damage occurred inside houses, and could have been reduced by installing temporary flood barriers, protections on basement light wells, anti-flood backflow valves on sewer connections, and by moving various in-house appliances like electricity, above the highest known water level. Many of these measures, which undoubtedly reduce the vulnerability of residential housing, were in fact mandated before the 2016 disaster: it is typically the case with sewer backflow valves which are mandated in the sewerage regulations, or with moving up electric appliances, mandated in most PPRi. However, as usual in France, regulations do not apply retrospectively so these rules apply primarily to new buildings and eventually to post-disaster reconstruction. Insurance companies are not obliged to fund additional costs, but they tend to do it more, for instance by paying for replacing wood floors with tiling at ground level.

Another type of measure indirectly supporting vulnerability control is the reduction of insurance coverage in the case of repeated disasters. As we wrote above, insurance companies were not allowed to raise premiums after a disaster. But this rule was indirectly relaxed when insurance and reinsurance companies realised they were compensating the same landowners in the same communes, for the same works several years in a row! With the reform adopted in 2000, where communes do not have a PPR, the (residential and business) deduction on insurance reimbursement[6] (*franchise*) is raised when events are repeated: since 2001, if in the last 5 years there were three events, the deductible ('excess' in the UK) is doubled, with 4 events it is trebled, and with 5 events it is quadrupled. In the period between 2000 and 2017, only 5% of contractual deductibles were thus modulated, and less than 1.5% were more than doubled. What then appears as a credible threat should incentivise the most vulnerable local authorities to adopt a PPRi faster, and in turn residents to invest in vulnerability reduction.

The most innovative measure was mentioned at the beginning of this article: the PAPI. PAPIs were initiated by communes and catchment institutions which realised that drafting the PPR did not in itself reduce the vulnerability of existing construction, and something more had to be done. For all types of flooding (river overflowing, groundwater rising, stormwater flooding and sea surge), local authorities are eligible to tap a special fund derived from Cat' Nat' called Barnier fund or *fonds de prévention des risques naturels majeurs* (Prevention Fund on Major Natural Risks – FPRNM) (Figure 2), to reduce vulnerability before a disaster takes place. This began when environment minister Michel Barnier sought a solution to move the population of a village before a neighbouring cliff would collapse on it. Since houses could not be sold, a solution had to be found to buy their property amicably and allow them to relocate. Several other cases were then identified, and finally in 2003 it was decided to call for tenders of local projects to reduce vulnerabilities; and a yearly percentage of the Cat' Nat' fund was diverted to

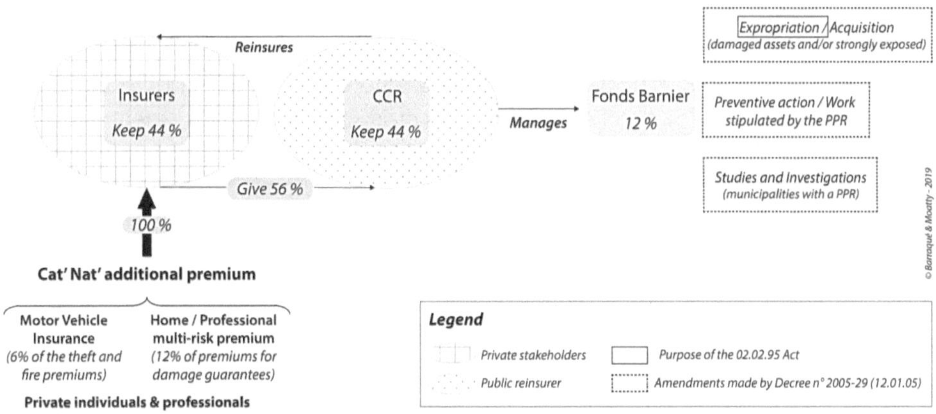

Figure 2. Barnier fund origin and eligible operations (authors' elaboration).

subsidise these projects. The call for tender was a success since more than 100 PAPI were developed, covering respectively 40% and 60% of the population exposed to risks from flooding and from sea surges.

In 2017, the transfer to the Barnier fund had reached 12% of the Cat' Nat' budget, *i.e.* providing 200 mln € at national level (representing 38% on average of the PAPI investments). Although insufficient compared to all the vulnerability reduction projects, as is now being revealed with the FD implementation, it is difficult to increase the funding from Cat' Nat' for fear that insurance companies would not be able to cover post-disaster damages, whether in case of a major event, or in case of cumulated events of lesser magnitude repeated over several years.

The most recent measure concerning resilience was a decision of the government in 2014 to decentralise the management of flood control infrastructure at local level. On the one hand it is a good decision to give local authorities the responsibility for both the structural measures and for land-use based solutions. But on the other hand, observers suspect that after the Xynthia disaster in 2010, the government wanted to put an end to its own responsibility for dyke maintenance, so as to reduce its deficit. Four new responsibilities were devolved to local authorities, together labelled *Gestion de l'Eau, des Milieux Aquatiques et Prévention des Inondations* (management of water and aquatic environment and flood prevention – GEMAPI). To allow them to face resulting financing needs, local authorities are allowed to tax local residents and landowners up to 40 €/cap/yr to cover this new expense. After 4 years on 1 January 2019, only one third of EPCI had adopted this new tax (ADCF, 2019), and with rates much below 40 €, since they are already under pressure from the national accounting office for overspending. On top of it, GEMAPI tax is set in part on local housing taxes which the government intends to phase out progressively, as part of general taxation reduction on modest revenues. So it remains to be seen how this further decentralisation move will improve resilience.

A local example

To illustrate this situation of limited progress, we now cover a case in an area south of Paris, where a small river called Yvette is a tributary to Orge river, which is in turn a tributary to the Seine upstream of Paris. Yvette suffered a serious flooding episode in May-June 2016, (CCR, 2016), which was eligible to Cat' Nat' funding.[7] But in fact, flooding had previously taken place repeatedly, in particular downstream where the Yvette and the Orge meet before merging into the Seine: 17% of the 121 communes in the catchment were supported by Cat' Nat' 9 times or more between 1983 and 2016! Saint-Rémy-les-Chevreuse, upstream on the Yvette, was also badly flooded several times, in particular in 2016 when a nearly 100 year flood occurred and flooded 300 homes, 10 public buildings and 6 shops. Little was done to stop this repeated flooding directly, while some wetland restoration recently improved the situation.

One has to remember here that the *agences de l'eau* can support the implementation of the programme of measures of the Water Framework Directive (WFD – 2000/60/EC), including measures on wetlands that have indirect, positive impacts on floods; but cannot support direct flood control measures in the flood risk management plans of the FD. In the case study, the area is partly covered by a regional natural park,[8] the institution of which funded important efforts towards sustainable landscape governance. The catchment institutions of the Orge and Yvette rivers' and the park together initiated a *Schéma d'Aménagement et de Gestion des Eaux* (catchment plan – SAGE) as soon as 1997. A *Commission Locale de l'Eau* (local water committee – CLE) was set up in 1999, under participative democracy principles: the committee includes 30 elected representatives from the (116) municipalities and the (two) counties, 24 user representatives (farming union, fishing, consumer and environment associations, a university, the regional park, etc.) and 9 representatives of the Government (regional and county prefects, the *agence de l'eau*, regional services of various ministries), etc. Once the SAGE was drafted and adopted by the CLE in 2006, and approved by the prefect, resulting investments were made by the local river institution which is competent on floods and also on waste water: a joint board acronymed SIAHVY.[9] The plan had to be updated following the water law of 2006, and the new SAGE was approved in 2014. After the 2016 flood, the decision was made to develop a PAPI to try reducing vulnerability.

In the SAGE approved in 2006, there were already three major targets dealing with floods: protection of people and goods in lower areas (Flood Risk Management Plan, integration in urban plans, flood expansion areas); protection of people and goods from flooding due to runoff (specific stormwater control measures); and developing a risk culture in the population. These projects were picked up again in the revised plan in 2014, and included in the integrated catchment planning carried out by the SIAHVY. Some projects were being prepared before the serious flooding of May 2016, and could be implemented rapidly. In particular, the re-naturalisation of a small tributary of Yvette, the Mérantaise, was carried out to accommodate and store flood waters in a rural site with no damage to properties. Purchase of 800 ha and restoration of the wetland, a 4.5 million € project, was eligible for funding from the Agence de l'Eau Seine Normandy, as part of the programme of measures of the WFD. This project could improve resilience directly, but also indirectly through improving the legitimacy of the PAPI project.

The Mérantaise operation was a success and is now a show case for the *agence de l'eau*, since in the 2018 flood, there was no damage in that area; and elected representatives are now better convinced of the merits of land-based flood control. The PAPI Orge-Yvette was then drafted: it contains 8 types of actions, covering different phases of risk management (prevention and crisis management) and targeting different scales, from the watershed to the citizen:

- Improving risk knowledge
- Improving citizens' consciousness of the risk
- Monitoring and forecast of flooding episodes
- Alert and crisis management
- Integration of risk maps in town plans
- Reducing vulnerability and improving resilience
- Channelling the high flows
- Maintenance of flood protection structures

In addition, a transversal axis for governance includes a web site to share information between the institutions. In terms of vulnerability reduction and resilience improvement, the chief action is to help the most impacted municipalities in being prepared: drafting vulnerability assessments on public buildings, workplaces and shops, and setting up a strategy to cope with dramatic episodes in real time and hasten recovery. The cumulated cost of the 35 forecast actions of the plan is estimated at 1.57 M€ plus VAT (Mérantaise operation excluded), two thirds being devoted to flood knowledge and consciousness, and 200,000 € to improving resilience. The expected PAPI funding would mobilise the Barnier fund for 45%, the *Agence de l'eau* for only 4%, the County council for 26%, with the rest being self-funded by the builders/managers of infrastructure.

However, local managers of the Yvette catchment are worried about the area being eligible for the Barnier fund, due to the small size of the issue compared to other cases in France. They fear delays in implementation due to bureaucratic control of their projects, leading in the end to a 'temporal dilution' of the initial good will of local authorities. Staff in charge of the PAPI are concerned that it does not include more precise anticipation measures to improve resilience in post-disaster recovery.

In addition, vulnerability reduction funding focuses almost entirely on public and business infrastructure, and not on housing: the staff in charge of the PAPI also fear that individual landowners at risk will not invest in vulnerability reduction due to lack of incentives. They propose residents install anti-backflow valves to prevent flooding from sewers, but it is not mandatory, and people are reluctant due to the high cost. In other words, residents are rich enough to be ineligible for benefits, but cannot afford to invest on additional resilience measures, which are not included in the Cat' Nat' indemnity. The worst case is when insurance companies dramatically increase the reimbursement deduction if vulnerability reduction investments are not made. An example was given of an insurance company threatening to raise the *franchise* substantially if buildings were not equipped with removable light walls to block ground floor entrances, and flood barriers. This may ultimately lead people to quit their insurance. Conversely, these requests are met by the managers of public buildings like schools despite the cost, as being part of the PAPI.

Concluding remarks

Post-disaster recovery can be either addressed by collective or individual coping strategies, planned or spontaneous, with varying objectives depending on the stakeholders.

The French Cat' Nat' system combines individual and collective coping; it clearly represents an improvement in terms of recovery. Its success is due to its funding by all policy holders, be they at risk or not. This globally allows supporting recovery of extreme events without subsequently raising the insurance premiums of the victims, which obviously limits the risk of policy holders exiting. It can be compared with the American National Flood Insurance Program (NFIP) which for Lamond and Penning Rowsell (2014) is a typical example of 'add-on or separate policy, state-backed': insurees are eligible for subsidised extra flood insurance in areas which are largely self-designed as hazard areas. This leaves large areas outside of the scheme, and in addition, the price of extra insurance may be too high, leading to some policy holders exiting and potential fund bankruptcy.

If we compare with Italy, which has one of the worst disaster vulnerabilities in Europe, there is a clear difference: there the risk map is made and updated at national level, but priorities to develop preventative measures are left to local authorities, which are often too small to really control land-use developments at risk. The focus is on emergency management, and not on mid- and long-term territorial resilience. At least in France, Cat' Nat' binds local authorities and the Government through the insurance system and the PPRi together.

However, it seems that Cat' Nat' does not really encourage vulnerability reduction, in particular as concerns private landowners. Considerable amounts of money are invested in recovery but most of them finance an 'identical' reconstruction. This is probably nothing new for flood insurance experts at international level. But we want to highlight the difficulty in bridging flood prevention and global river and aquatic environment policy (FD and WFD directives' implementation). This is partly due to the ongoing economic crisis and the excessive national deficit compared to the Maastricht 3% target. Thus, in line with the spirit of the decentralisation laws, the concern of the present Government is to transfer its previous responsibilities on to local authorities. But they also want to regulate the use of public funds to make sure that they are not used, after disasters, to make up for the maladaptation or even the lack of maintenance of public assets before disasters (*e.g.* dykes, the responsibility of which falls to local authorities since 2014); and that this support does not increase the pre-existing vulnerability. In addition, Cat' Nat' reimbursements are reduced if floods occur repeatedly in an area in which stakeholders have not taken any preventive or mitigation measures. Facing this relatively negative / punitive approach to flood control, integrated river policies at catchment scale frequently include more positive measures to improve social and territorial resilience. But they also frequently stop at the gate of private properties.

This can be illustrated with the difficulties in drafting the PPRi. They were first decentralised to municipalities, but soon recentralised in the prefectoral services. So decentralisation was only partial, with local elected representatives mainly continuing to support the urban development potential of flood prone areas at the expense of vulnerability reduction. Resilience anticipation is thus limited by this typical characteristic of French political system: the State is above the citizens but private property is well defended. While landowners often manage to undermine the building bans or limitations in PPRi, the State is tempted to underestimate the impact of floods on private property and the need for adaptations, or leave it to the insurance companies.

Of course, it is much easier to draft a more severe PPRi and a more resilient land-use plan in the aftermath of a disaster: in La Faute-sur-Mer today, in the place of the housing estates where 29 people drowned in the February 2010 Xynthia event, there is now a golf course. Yet anybody can ask why wasn't it zoned that way beforehand?

One answer is to reconsider land-use planning from the perspective of the aquatic environment, and derive planning rules from a catchment plan, which is elaborated more in terms of water usership than of landownership. Participative democracy, which was particularly developed in the water sector in the last 50 years, may then help mitigate the well-known State *vs* private property French antagonism. In places where local stakeholders convene to draft a SAGE (catchment plan) there is a better potential involvement of the population; yet flood control measures directly concerning landowners as such are not frequent, as our case study illustrates. This is also because in France, catchment institutions and the *agences de l'eau* are not encouraged to include flood control measures in their tasks: they would need to get a corresponding funding mechanism beyond what is made available through the Barnier fund, which is reckoned as largely insufficient to trigger improved resilience in most vulnerable areas, while remaining complex to obtain.

Insurance companies are aware of the difficulties encountered after floods by impacted landowners. Yet they have to admit that, paradoxically, the Barnier fund remains partly unspent year after year, and unfortunately any positive fiscal cash flow attracts the envy of the Treasury! After taking 10% of the 200 million € since 2016, this year instead the total transfer from the Cat' Nat' fund is capped at 137 million €. Facing this unsatisfactory bottleneck, the Government and the CCR (2016) are considering an evolution of the regime, at least with two measures: subsidising temporary relocation of victims, since recovery/reconstruction duration is frequently much longer than expected; and reimbursing some additional costs corresponding to vulnerability reduction in the damages payment calculations by insurance companies.

But again this raises the general question of the trade-off between after and before disasters particularly in Cat' Nat': the more one spends on more resilient recovery, the more is needed for the PAPI and Barnier funding, and the less is left to cover disasters after they occurred. And all this in a context where climate change increases the uncertainty on financial sustainability, not only for flooding but also for drought events which seem to multiply in France.

Some observers consider it would be a good idea to transfer the Barnier fund to the 6 *agences de l'eau*, which could then merge this budget with their own budget aimed at recovering the good status of the aquatic environment, and help develop more integrated WFD and FD policies. This would be a reversal of the previous policy leaving the *agences* out of the issue of flood control. But the way chosen in 2014 is somewhat different: consolidated local authorities are allowed to raise a new tax on landowners, the above mentioned GEMAPI. It remains to be seen how the new legal responsibility on flood planning and infrastructure devolved to local authorities will help boost this important but yet fragile task: a more resilient recovery thanks to more active policies on land-use planning and on buildings restoration.

In the Netherlands, after the 1953 disaster (1830 casualties), insurance companies refused to cover the flood risks, and then the government developed a strong collective protection system based on dykes at two levels (national and regional). Today most of the flood control policy is implemented at the level of the former polder waterboards, which were dramatically consolidated since this event into what they call Regional Water Authorities (RWAs). But many people think that this focus on flood protection through structural measures has been overly disconnected from land-use management by local authorities, resulting in an aggravation of vulnerability (Doorn-Hoekveld, 2018). So they tend to promote a new 'multi-layered' policy, where improvement of the aquatic environment would be combined with alternative flood control policies: returning space to the rivers, mobilising private land to temporarily store and discharge excess water (with due compensation of the losses); and above all, developing better links between RWAs, and provinces and local authorities in charge of land-use management so that they are incited to reduce spatial vulnerability. Local zoning plans and some building permits are now subjected to a 'water test' to clearly inform local authorities of their consequences in terms of flooding. This new 'living with water' approach even includes an experiment on insuring some assets against floods: the 'Neerlandse' insurance was initiated in 2011 but has not attracted many subscribers yet.

If the French would draw on the Dutch multi-layered policy, they could support a more systematic co-ordination between the consolidated local authorities (EPCI) and the catchment institutions (EPTB), so as to plan vulnerability reduction through land-use control, including with compensated storage of excess water on private property; and funding could combine the Barnier fund, the GEMAPI tax levied by the EPCI, plus some specific mutual levy system to be developed at the EPTB level, in the aim at better bridging the implementation of the FD and the WFD.

Notes

1. Act n°82-600 of 13 July 1982 for the compensation of victims of natural disasters. Floods represent more than 90% of events.
2. Directive 2007/60/EC of the European Parliament and of the Council of 23 October 2007 on the assessment and management of flood risks.
3. Quoted from the 1982 Cat' Nat' law. Our translation.
4. The FSCT is a financial package announced in the Decree in the Official Journal of 27 August 2008. So far, we have limited information on its performance.
5. *Act n° 2004–811 of 13 August 2004 dite « loi de modernisation de la sécurité civile » : civil security modernisation law, art. 13.*
6. In some insurance contracts there is a fixed deductible on reimbursement, in exchange with a premium reduction.
7. Severe storms followed by an extreme rainfall episode provoked floods in several parts of France between May 25 and June 6, resulting in damages on insured properties up to 1.4 bn €
8. Parc Naturel Régional de la Haute Vallée de Chevreuse. https://www.parc-naturel-chevreuse.fr/
9. SIAHVY : Syndicat Intercommunal pour l'Aménagement Hydraulique de la Vallée de l'Yvette

Disclosure statement

No potential conflict of interest was reported by the authors.

References

Association des Communautés de France (AdCF). (2019, February 8). Gemapi: 35% des communautés ont désormais institué la taxe affectée. In *AdCF-direct*, Newsletter n°457. Retrieved from https://

www.adcf.org/contenu-article-adcf-direct?num_article=4533&num_thematique=1&id_newsletter=382&source_newsletter=457&u=MTc3Ng

Barraqué, B. (2014). The common property issue in flood control through land use in France. *Journal of Flood Risk Management*. doi:10.1111/jfr3.12092. (Virtual Special Issue: *The European Flood Risk Management Plan*, Guest Editors: Asst. Prof. Dr. Thomas Hartmann and Prof. Dr. Robert Juepner).

Bourrelier, P. H. (Dir.) (1997). *La prévention des risques naturels : rapport de l'instance d'évaluation*. Paris: la Documentation Française, 702 p.

Caisse Centrale de Réassurance (CCR). (2018). *Les Catastrophes naturelles, en France, bilan 1982–2018*. Self-Published Report. Retrieved from http://www.side.developpement-durable.gouv.fr/EXPLOITATION/ACCIDR/doc/IFD/IFD_REFDOC_0550808

CCR – Service R&D modélisation – Direction des Réassurances & Fonds Publics. (2016). *Inondations de mai-juin 2016 en France. Modélisation de l'aléa et et des dommages*. Report. Retrieved from https://www.ccr.fr/documents/23509/29230/Inondations+de+Seine+et+Loire+mai+2016+avec+Prevention_28062016.pdf/b3456170-8dee-4c3e-b2aa-608172cce8c7

Christoplos, I. (2006, Février 2–3). *The elusive "window of opportunity" for risk reduction in post-disaster recovery*. Intervention in ProVention Consortium Forum, Bangkok, 4 p.

Doorn-Hoekveld, W. J. (2018). *Distributional effects of EU flood risk management and the Law. The Netherlands, Flanders and France as case studies* (PhD thesis). Utrecht University, Utrecht, 231 p.

Duran, P., & Thoenig, J. C. (1996). L'Etat et la gestion publique territoriale. *Revue Française de Science Politique*, *46*(4), 580–623.

Grémion, P. (1976). *Le pouvoir périphérique, bureaucrates et notables dans le système politique français*. Paris: le Seuil, 477 p.

Lamond, J., & Penning Rowsell, E. (2014). The robustness of flood insurance regimes given changing risk resulting from climate change. *Climate Risk Management*, *2*, 1–10.

Lamond, J., & Proverbs, D. G. (2009). Resilience to flooding: Lessons from international comparison. *Proceedings of the Institution of Civil Engineers - Urban Design and Planning*, *162*, 63–70.

Lamond, J., Proverbs, D. G., & Hammonds, F. N. (2009). Accessibility of flood risk insurance in the UK: Confusion, competition and complacency. *Journal of Risk Research*, *12*(6), 825–841.

Ledoux, B. (2000). *Retour d'expérience sur la gestion post-catastrophe dans les départements de l'Aude et du Tarn*. Rapport pour le Ministère de l'Aménagement du Territoire et de l'Environnement (MEDD, ex-MATE), Bruno Ledoux Consultants, 70 p.

Lefrou, C., Martin, X., Labarthe, J. P., Varret, J., Mazière, B., Tordjeman, R., & Feunteun, R. (2000). Les crues des 12, 13 et 14 novembre 1999 dans les départements de l'Aude, de l'Hérault, des Pyrénées-Orientales et du Tarn, Rapport au Ministre de l'Aménagement du Territoire et de l'Environnement, 99 p. + annexes.

Maujean, G. (2018, June 25). Pourquoi il faut changer le régime des Cat' Nat'. *Les Echos*. Retrieved from https://www.lesechos.fr/25/06/2018/lesechos.fr/0301876453903_pourquoi-il-faut-changer-le-regime-des—cat-nat-.htm

Mechler, R., Linnerooth-Bayer, J., Hochrainer, S., & Pflug, G. (2006). Assessing financial vulnerability and coping capacity: The IIASA CatSim model. Concepts and methods. In J. Birkmann (Ed.), *Measuring vulnerability and coping capacity to hazards of natural origin* (pp. 380–398). Tokyo: United Nations University Press.

Mission Risques Naturels (MRN). (2018). *Inondations Seine-Loire 2046. Etude sur l'endommagement du bâti causé par un événement 'inondation'*. Case Study Report. Retrieved from https://www.mrn.asso.fr/wp-content/uploads/2018/04/04-06-2018-rapport-mrn_inondations-seine-loire-2016.pdf

Moatty, A. (2017). *Post-flood recovery: An opportunity for disaster risk reduction?* In F. Vinet (Ed.), *Floods* (vol. 2, pp. 349–364). ISTE Press, Elsevier.

Moatty, A., Gaillard, J.-C., & Vinet, F. (2017). Du désastre au développement: Les enjeux de la reconstruction post-catastrophe. *Annales de géographie*, 714, Armand Colin, 169–194.

Moatty, A., & Vinet, F. (2016). *Post disaster recovery: The challenge of anticipation*. Web conference FLOOD risk 2016 - 3rd European Conference on Flood Risk Management. doi:10.1051/e3sconf/20160717003

Moatty, A., Vinet, F., Defossez, S., Cherel, J. P., & Grelot, F. (2018). Intégrer une éthique préventive dans le processus de reconstruction post-catastrophe: place des concepts de résilience et d'adaptation dans la reconstruction préventive. *La Houille Blanche*, SHF ed. ISSN: 0018-6368, 11–19.

MRN & et al. (2017, September 8). Conférence B3: exemples d'apports des acteurs de l'assurance française au 'faire et reconstruire mieux'. Retrieved from https://www.mrn.asso.fr/wp-content/uploads/2017/12/2017-presentation-apres-midi-mrn-aqc-elex-saretec-conference-b3.pdf

Quarantelli, E. L. (1999). *The disaster recovery process: What we know and do not know from research.* Preliminary paper n°286. Newark, DE: University of Delaware, Disaster Research Center.

United Nations International Strategy for Disaster Reduction (UNISDR). (2013). *Global assessment report on disaster risk reduction. From shared risk to shared value: The business case for disaster risk reduction*. 288 p.

Vinet, F., Boissier, L., & Defossez, S. (2011). *La mortalité comme expression de la vulnérabilité humaine face aux catastrophes naturelles : deux inondations récentes en France (Xynthia, var, 2010). VertigO* [Online], Volume 11 Numéro 2 | septembre 2011, consulté le 01 août 2018; doi:10.4000/vertigo.11074.

Worms, J. P. (1966, July-September 8–3). Le préfet et ses notables. In *Sociologie du Travail* (pp. 249–275). Paris: Le Seuil.

OPEN ACCESS

An assessment of best practices of extreme weather insurance and directions for a more resilient society

P. Hudson, L.T. De Ruig, M.C. de Ruiter, O.J. Kuik, W.J.W. Botzen, X. Le Den, M. Persson, A. Benoist and C.N. Nielsen

ABSTRACT

Extreme weather resilience has been defined as being based on three pillars: resistance (the ability to lower impacts), recovery (the ability to bounce back), and adaptive capacity (the ability to learn and improve). These resilience pillars are important both before and after the occurrence of extreme weather events. Extreme weather insurance can influence these pillars of resilience depending on how particular insurance mechanisms are structured. We explore how the lessons learnt from the current best insurance practices can improve resilience to extreme weather events. We employ an extensive inventory of private property and agricultural crop insurance mechanisms to conduct a multi-criteria analysis of insurance market outcomes. We draw conclusions regarding the patterns in the best practice from six European countries to increase resilience. We suggest that requirements to buy a bundle extreme weather event insurance with general insurance packages are strengthened and supported with structures to financing losses through public-private partnerships. Moreover, support for low income households through income vouchers could be provided. Similarly, for the agricultural sector we propose moving towards comprehensive crop yield insurance linked to general agricultural subsidies. In both cases a nationally representative body can coordinate the various stakeholders into acting in concert.

1. Introduction

Extreme weather events can have large impacts on society. For example, in 2016 a combination of flash floods and storms in Germany, Belgium, and Switzerland inflicted \$2.2 billion in losses. Similarly, across central Europe a hail storm led to \$1.9 billion in damage (Munich Re, 2017). About 50% of these costs were absorbed by insurers. However, this is not always the case. There can be limits to insurance, as can be seen

Supplemental data for this article can be accessed at https://doi.org/10.1080/17477891.2019.1608148.

when insurers leave markets or restrict coverage (Cremades et al., 2018). A sufficiently resilient society can limit these impacts, as resilience is a proactive and forward-looking concept. While there is no commonly accepted definition of resilience, resilience can build upon three pillars: resistance, recovery, and adaptive capacity (Thieken, Mariani, Longfield, & Vanneuville, 2014). Resistance is the ability to limit disaster impacts. Recovery relates to the time needed to return to the pre-disaster state. Adaptive capacity refers to the ability to be better prepared for future disasters.

There are several strategies for improving the resistance and recovery pillars. These strategies can be broadly grouped into prevention, mitigation, and risk-transfer. Prevention measures are put in place to prevent damage from certain extreme weather events from occurring. One example of such a measure is a dike (Aerts, Botzen, De Moel, & Bowman, 2013). Mitigation actions limit damage when an event occurs, for example, building codes (Burby, 2001) or property-level measures (Kreibich, Thieken, Petrow, Müller, & Merz, 2005) that reduce the susceptibility to damage. Risk-transfer strategies, on the other hand, do not decrease the direct impacts from an event. Instead, they allow financial reserves to be developed in order to aid recovery, thereby helping people get back on their feet (Botzen, 2013). Additionally, as prevention and mitigation cannot prevent all of the potential impacts of extreme weather events, risk-transfer helps society to manage the remaining risk, if risk-transfer and mitigation are sufficiently linked, allowing for proactive risk management.

Proactive risk management is required to minimise current and future extreme weather impacts. The Sendai framework for disaster risk reduction for the period 2015–2030 prioritises developing societal resilience through the use of measures that finance recovery costs while incentivising risk reduction (Mysiak, Surminski, Thieken, Mechler, & Aerts, 2016). Insurance, the prime example of risk-transfer, fills these roles by providing compensation after an extreme weather event, while acting as a potential price signal of risk. A strong price signal provides an incentive for active policyholder-level risk management. Additionally, insurance regulations or requirements can enforce a certain low level of vulnerability before potential policyholders can be insured. Finally, a functioning insurance sector can aggregate and disseminate information before and after disasters.

However, despite the potential for insurance to act as a transformative strategy for meeting the objectives of the Sendai framework and adapting to climate change, the role of insurance is viewed differently across stakeholders, depending on their level of risk and cultural context. Acknowledging the importance of local context is important, as it shows that there is not a one-size-fits-all solution – what might be acceptable in one region may not be in another, leading to the idiosyncratic development of insurance markets (Surminski et al., 2015). These differences can allow for a range of structures and outcomes that can be studied, and potential patterns in success factors and common outcomes can be identified.

To this end, we conducted a comparative study of private property and agricultural insurance in 12 different countries for a range of extreme weather events. The objective was to identify best practice in order to draw conclusions about improving extreme weather resilience, which can inform future policy directions across Europe. For private property insurance, we investigated floods (pluvial, fluvial, and coastal), hail, and windstorms, and for the agricultural insurance sector, we also included droughts.[1] We developed a holistic evaluation framework based on a multi-criteria analysis (MCA) in order to judge best practice holistically across extreme weather events and five key outcomes: insurance penetration rates, risk-signalling and risk-reduction incentives, insurance affordability and availability, the speed and certainty of payments, and insurer solvency. These outcomes were determined by a review of the literature and stakeholder engagement. Moreover, we assumed the presence of three different risk management objectives: providing high degrees of coverage and affordability; balancing the different objectives of relevant stakeholders; and the ability of insurance to act as a risk management mechanism.

One novel way that we extend the current literature is by collecting and reporting new information on the functioning of insurance markets for extreme weather risks in Europe holistically across multiple hazards, rather than focussing on a single risk, as is common in the academic literature, e.g. Michel-Kerjan and Kunreuther (2011), Hudson, Botzen, Feyen, and Aerts (2016) or Unterberger, Hudson, Botzen, Schroeer, and Steininger (2019). On the basis of this data collection, we draw conclusions about best practices for Europe that can be used to enhance the flood resilience of European households. Additionally, our study extends the nascent literature investigating insurance mechanisms via MCA. For example, Hochrainer-Stigler and Lorant (2018) who presented an MCA of potential reforms to the European Union Solidarity Fund or Unterberger et al. (2019) who study potential reforms to the Austria Catastrophe Fund. We develop this limited literature by presenting an MCA framework and data collection approach that was used to study a range of extreme weather events across varying contexts within 12 European countries, across two insurance sectors.

In conducting the MCA, we identified Denmark, France, Spain, and the UK as exemplars of best practice for private property insurance, while for the agriculture sector the exemplars were Austria, Spain, and Sweden. Across both sectors we find that, despite the differing development trajectories for insurance markets, the best practice cases display similar patterns of behaviour even if the implementation differed in practice. Therefore, our results indicate a set of market features that improve flood resilience if applied to less well-performing markets. Moreover, the relatively low scores achieved in the MCA indicate that there is room for improvement in how Europe uses insurance to boost flood resilience, by focusing more on the resistance and adaptive capacity-boosting elements of resilience.

Finally, the suggested series of policy recommendations for increasing resilience has features similar to those of the reforms for the National Flood Insurance Program in the United States as proposed, for example, by Michel-Kerjan and Kunreuther (2011). This demonstrates the overall generalizability of both these results and policy directions at the global level regarding how insurance could move forwards.

2. Methodology

2.1. The interaction of insurance and resilience

The traditional role of insurance is focused upon enhancing the recovery pillar of resilience, with insurance providing financial protection against extreme weather events. Insurance coverage replaces a randomly occurring large loss with a smaller fixed cost, which is welfare enhancing for risk-averse individuals (Mas-Colell, 1995). The rapid provision of compensation payments is especially important in the wake of large events. For example, the 2002 flood in Germany caused an average loss of €58,000, while the 2013/2014 winter flooding in the UK inflicted an average loss of €35,000 (Chatterton et al., 2016). Such amounts can be beyond the means of those affected to be able recover from the disaster in a timely manner. The provision of secure compensation limits the long-run negative impacts of a slower recovery (Botzen, 2013; Schäfer, Warner, & Kreft, 2019). Therefore, sufficient insurance coverage improves the recovery pillar by increasing and diversifying the policyholder's recovery capacity.

Concerning the resistance pillar, insurance can also play a role in proactively limiting impacts before extreme weather events occur (Kunreuther & Michel-Kerjan, 2009). One such avenue is the provision of direct incentives for policyholder risk reduction. For example, by allowing premiums to reflect the underlying risk, policyholders can be provided with premium discounts if they implement risk-reducing measures (Hudson et al., 2016; Kunreuther & Michel-Kerjan, 2013). The second is in setting minimum requirements in order to be insured, such as meeting building code regulations, which has been successful in the USA (Burby, 2001).

The final pillar is adaptive capacity. A functioning insurance market can contribute towards creating a more adaptively capable society by acting as an information generation and dissemination platform. For example, insurers require diverse portfolios of policyholders who undertake different strategies for extreme weather event risk management. The insurer is in a position to aggregate these experiences and see which measures are more likely to be successful and can share this information with policyholders. Additionally, a functional insurance market can facilitate adaptive capacity, as the various risk management stakeholders contribute different areas of expertise. This combination of expertise creates a more detailed understanding of where and how the disaster impacts materialise, which would not be available if stakeholders were not encouraged to act in concert.

2.2. Case study selection

This section is a select summary of the characteristics of insurance mechanisms, the details of which we reported in our report published by the European Comission (2017), which contains the underlying data and inventory of insurance mechanisms. In that report, we used the inventory proposed in Paudel et al. (2012) as a baseline for data collection. The table and adapted definitions from the underlying report are presented in Appendix 2. We reviewed the academic and grey literature to fill in the requirements of the inventory. An initial literature search was conducted for the European Union. The final 12 case studies shown in Figure 1(a) were selected based on data availability, quality, and consistency for the studied insurance markets. Following this selection, relevant local stakeholders (e.g. national insurance associations) were approached to assess the validity of this information and adjust and update it as required.

Figure 1(b and c) provide examples of the information uncovered. They highlight the degree to which the public and private sectors collaborate with regard to extreme weather insurance. The information indicates that the majority of extreme weather insurance is provided by the private market. The remainder of the extreme weather insurance is provided through a combination of public-private partnerships. These public-private partnerships can vary in scope across countries. For example, the French partnership covers a

Figure 1. Selected European case study countries, noted in dark blue.

large range of extreme weather events, while the UK partnership is focused upon flooding. However, there is no country that relies solely on publicly provided insurance, even though public compensation schemes exist.

In terms of provision, in the private property sector wind and hail storms are generally covered as part of household insurance policies. Flooding coverage tends to be voluntary, which is associated with low coverage rates. In the agricultural insurance sector, the event most commonly protected against is hail.

2.3. Extreme weather event insurance assessment criteria

Functional extreme weather event insurance markets must meet conditions of actuarial and economic insurability (Charpentier, 2008). Actuarial insurability can be defined as consisting of random quantifiable events with losses being relatively uncorrelated across policyholders. Economic insurability requires the absence of major information asymmetries between market agents and that there be sufficient overlap between the consumers' willingness to pay for insurance and the offered premiums. However, meeting these conditions is not sufficient to determine best practice, as meeting the insurability conditions produces a range of trade-offs. For instance, there is a known trade-off between the incentive for risk reduction and the affordability of premiums. Additionally, the importance of these trade-offs differs across stakeholders and their objectives within the public policy process (Surminski, 2018).

Insurance market outcomes should be evaluated within their wider contexts. We base the best practice benchmarks on a set of criteria drawn from risk management objectives for insurance from the academic and grey literature, for example, which is detailed in Appendix 1: Golnaraghi, Surminski, and Schanz (2017), Kunreuther (2017), Surminski (2018), The Geneva Association (2018). Additionally, during the completion of our report we consulted expert stakeholders from the insurance industry and academia via three project workshops/events, a series of email/telephone interviews, and a review of the public consultation responses to the European Commission Green Paper on the Insurance of Natural and Man-made Disasters. This process highlighted the importance of the following outcomes: the overall penetration rate across extreme weather events, risk signalling or risk reduction ability, the ability to absorb large losses, the ability to provide quick and certain compensation, and the overall affordability and availability of insurance. See Appendix 1 for details.

2.4. Extreme weather event insurance evaluation criteria

For each of the criteria, a market sector is awarded points according to Table 1. Moreover, while the core evaluation criteria are the same across both investigated sectors, the criteria are operationalised differently. The criteria presented in Table 1 were presented twice to groups of stakeholders in order to receive feedback on the overall suitability of the points scoring system. The stakeholders broadly accepted the scoring system, with minor suggestions for use in the developed MCA framework. Appendix 2 provides a more detailed description of the points scoring system.

The MCA expresses a sector's performance with a single overall score, while allowing each outcome to be taken into account according to its perceived importance. This is shown in Equation 1 as the overall score equalling the weighted sum of the individual standardised criterion scores:

$$S_{c,s}^1 = \sum_{i=1}^{i=5} \omega_i^1 \mathrm{MCA}_{c,s,i} \tag{1}$$

$S_{c,s}^1$ is the overall score for country c and sector s under criterion Weighting Scheme 1 (superscript 1). This score is the weighted sum of the individual standardised criterion scores (i) for country $c(\mathrm{MCA}_{c,i})$. The criterion weights (ω_i^1) are defined per weighting scheme as a proxy for risk management objectives. The value for $\mathrm{MCA}_{c,s,i}$ is produced by following either Equation 2, for a continuous ranking, or Equation 3, for a relative ranking.

$$\mathrm{MCA}_{c,s,i} = \frac{(\mathrm{MCA}_{c,s,i})^{\text{raw score}}}{\text{Max points possible for criterion } c} \tag{2}$$

$$\mathrm{MCA}_{c,s,i} = \frac{(\mathrm{MCA}_{c,s,i})^{\text{raw score}} - (\mathrm{MCA}_{c,s,\min})^{\text{raw score}}}{(\mathrm{MCA}_{c,s,\max})^{\text{raw score}} - (\mathrm{MCA}_{c,s,\min})^{\text{raw score}}} \tag{3}$$

The weights are based on one of three scenarios (see Table 2): Weighting Scheme 1: Providing high degrees of coverage and affordability; Weighting Scheme 2: Balancing the different objectives of relevant stakeholders; and Weighting Scheme 3: The ability of insurance to act as a risk management mechanism.

Weighting Scheme 2 was developed based on input received from a series of stakeholder questionnaires (see Appendix 1). One aspect of these questionnaires was to ask the respondent to rank the five criteria in terms of their importance. However, the responses yielded similar weights, as most completed responses were provided by insurers and academics, which limited the scheme's overall representativeness. Weighting Schemes 1 and 3 were developed based on the judgement of the research team in order to accommodate the different risk management objectives that were retrieved from the literature review and stakeholder engagement process.

As an illustration, take Spanish private property insurance. For Criterion 1, the median insurance penetration rate was ~90%, which scores 4 points. For Criterion 2, the sector relies on small deductibles with flat rate premiums, which earns 1 point (if risk-based premiums were used, Spain would have scored 3 points). For Criterion 4a, all the relevant hazards are covered by the Spanish insurance system, scoring 3 points. For Criterion 4b, due to the income distribution and the average property insurance premium the rate of

Table 1. Scoring metric the private property insurance sector.

Private property insurance							Agriculture (crop) insurance						
		4 points	3 points	2 points	1 point	0 points			4 points	3 points	2 points	1 point	0 points
Criterion 1: Insurance penetration rate	The percentage of households that have coverage against the set of relevant extreme	[81,100]	[61,80]	[41,60]	[21,40]	[0,20]	Criterion 1: Insurance penetration rate	This is measured as the percentage of arable land that is insured against the set of relevant extreme weather events.	[40+]	[31,40]	[21,30]	[11,20]	[0,10]
Criterion 2: Risk signalling ability	The ability of insurance market structures to act as a signal of risk or risk reduction potential			Large deductibles (i.e. larger than €2000); Premium discounts; Required vulnerability standards; Risk based premiums	Small deductibles (i.e. smaller than €2000 but larger than €0); Awareness campaigns	No risk signalling	Criterion 2: Risk signalling ability	This criterion measures the 'holistic' ability of insurance market structures to act as a signal of risk or risk reduction potential.			Large deductibles (i.e. a loss of 30% or more); Premium discounts; Required vulnerability standards; Risk based premiums; Bonus-malus	Small deductibles (i.e. a loss less than 30%); Awareness campaigns	No risk signalling
Criterion 3: Ability to absorb large losses	This criterion judges the ability of the sector to absorb large losses in the case of large scale events.		Unlimited State grantee for extreme weather event losses	Access to private reinsurance for all risks; Access to sector/ disaster wide pool for all risks; Access to public sector reinsurance for all risks; Limited State grantee for extreme weather event losses	Access to private reinsurance for some risks Access to sector/disaster wide pool for some risks Access to public sector Reinsurance for some risks	Reliance on internal capital only	Criterion 3: Ability to absorb large losses	This criterion judges the ability of the sector to absorb large losses in the case of large scale events.			Unlimited State grantee for extreme weather event losses	Access to private reinsurance for all risks; Access to sector/ disaster wide pool for all risks; Access to public sector Reinsurance for all risks; Limited State grantee for extreme weather event losses	Access to private reinsurance for some risks; Access to sector/ disaster wide pool for some risks; Access to public sector Reinsurance for some risks

(*Continued*)

Table 1. Continued.

		Private property insurance							Agriculture (crop) insurance				
		4 points	3 points	2 points	1 point	0 points			4 points	3 points	2 points	1 point	0 points
Criterion 4a: Affordability	The percentage of the population finding insurance unaffordable	[0,5]	[6,10]	[11,15]	[16,20]	[≥21]	Criterion 4a: Affordability	The premium as a percentage of the total insured value	[0,1]	[1,2]	[2,4]	[4,6]	[≥6]
Criterion 4b: Availability	The number of extreme weather events that can be insured against				1 point per extreme weather event that is insurable		Criterion 4b: Availability	The number of extreme weather events that can be insured against				1 point per extreme weather	
Criterion 5a: Quick and certain compensation payments	The certainty of receiving accurate compensation			No ad-hoc government compensation is possible. There are only formal mechanisms for risk-transfer with clear rules.	Limited private sector insurance coverage and ad-hoc government compensation; Specialist loss adjusters are used to assess and process claims.	Complete reliance on ad-hoc government compensation.	Criterion 5a: Quick and certain compensation payments	Certainty			No ad-hoc government compensation is possible. There are only formal mechanisms for risk-transfer with clear rules	Limited private sector insurance coverage and ad-hoc government compensation; Specialist loss adjusters are used to assess and process claims.	Complete reliance on ad-hoc government compensation
Criterion 5b: Quick and certain compensation payments Only information was that the claim was to be settled after harvest	*Speed*	*Not studied due to limited information*					Criterion 5b: Quick and certain compensation payments	Speed	Up to 15	days	Up to 30 days	Up to 60 days	Up to 90 days

Table 2. Selected weighting schemes used.

	Weighting scheme 1	Weighting scheme 2	Weighting scheme 3
Insurance penetration rate	0.35	0.23	0.125
Risk signalling	0	0.22	0.5
Ability to absorb large losses	0.15	0.19	0.125
Affordability and availability	0.35	0.19	0.125
Quick and certain compensation	0.15	0.18	0.125

unaffordability earns Spain a score of 0 points. Therefore, for Criterion 4 Spain has a total score of 3 points. Finally, regarding the speed of certainty and payments, Spain has formal systems with clear rules and no ad hoc payments, with the common usage of professional loss adjusters, resulting in 3 points. Overall, once standardised Spain has a score of about 0.78 when Weighting Scheme 1 is employed and 0.48 when Weighting Scheme 3 is used.

A country is declared to exemplify best practice if it was ranked first or second under either of the ranking methods for a given ranking scheme. We then conducted a qualitative analysis of the identified market sectors to see what lessons can be drawn by detecting common behavioural patterns.

3. Results and discussion

Figure 2 presents the MCA outcome by highlighting the best practice cases. See Appendix 3 for a more detailed description. It must be noted that, due to our methodological approach, this analysis is conducted from a policymaker perspective.

Table 3 presents a summary of the common features of the examples of best practice, which can help to identify ways to enhance societal extreme weather resilience. However, while a comparative analysis reveals general patterns which should be tailored to meet specific local conditions, the general patterns reveal a starting point. While this study is focused on Europe and draws lessons mainly from the practices within Europe, countries

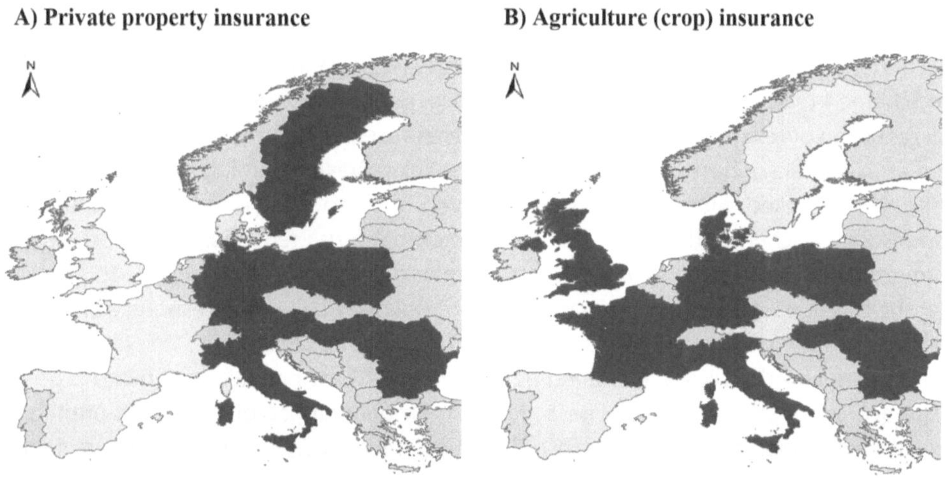

Figure 2. Countries in yellow are ranked as first or second out of the countries studied for at least one of the three weighting schemes.

Table 3. Summary of common features in the insurance markets studied.

Private property	Agriculture
Multiple extreme weather events are combined in a single policy.	The use of multi-risk, or yield, insurance.
The purchase of extreme weather event insurance is connected to a far more commonly required and enforced product.	All cultivated land must be insured.
Collaboration between public and private sector with a commonly stated and understood objective.	Premium subsidies direct investment to multi-risk policies.
Provision of a national pool or public reinsurance/support for catastrophic losses in addition to private reinsurance coverage.	Pool like structures or public reinsurance for systemic risks such as droughts.
	A tradition of collaboration between the public and private sector risk managers.

outside of Europe may also score highly in the MCA. For example, Australia displays many of the characteristics mentioned in Table 3, such as risk-based premiums and multi-peril coverage across a range of natural hazard insurance (Mcaneney, Mcaneney, Musulin, Walker, & Crompton, 2016).

An overview of the MCA results reveals the importance of flooding in driving the MCA results, as compared to windstorm or hail risks. Flooding is problematic because floods are highly localised, which limits the ease with which losses can be spread geographically and mutualised, in contrast to more spatially diverse hazards such as windstorms or hail. Overall, the average MCA score is highest under Weighting Scheme 1 and lowest under Weighting Scheme 3. Additionally, the standard deviation of the MCA scores fell moving from Scheme 1 to Scheme 3, indicating smaller differences in overall performance across countries. This indicates that, on the whole, the case studies are relatively successful at providing insurance as a social good. For private property insurance, the two main (and interconnected) problems to overcome are insuring flooding and encouraging households to buy insurance. This is due to the low rates of coverage outside of the best practice countries studied.

3.1. Resistance

3.1.1. Outcome of the MCA

The MCA results indicate that all 12 case studies score low regarding risk reduction, which is especially highlighted when the average score under Weighting Scheme 1 is compared to that under Weighting Scheme 3. Moreover, the standard deviation of scores falls from Weighting Scheme 1 to Scheme 3, indicating a more similar performance of countries as we focus on risk reduction or resistance-building aspects. Therefore, even in terms of the countries identified as exemplars of best practice, the resistance link is weak regarding how insurance and resilience are linked. Moreover, previous research has also established that is difficult to find insurance schemes that are successful at promoting risk reduction (Linnerooth-Bayer, Surminski, Bouwer, Noy, & Mechler, 2019; Surminski & Oramas-Dorta, 2014). This implies that the countries isolated as best practice are not isolated due to their resistance improving ability, but rather because they over-performed on the other criteria.

Denmark and the UK are identified as exemplars of best practice for private property insurance under Weighting Scheme 3 due to their general use of risk-based premiums, use of deductibles and a number of adaptation initiatives. French insurers are legally required to include a deductible starting at €380 and rising to €3,050 in line with the occurrence of floods. In these countries, the main source for providing resistance is the government rather than the policyholder. For example, in France local governments generate risk prevention plans, which can mandate certain risk reduction activities or requirements to be enforced in selected areas. However, the extent to which these plans can require risk-reduction measures is limited in practice (Poussin, Botzen, & Aerts, 2013). Spanish property insurance also does not provide signals for risk reduction because risk reduction is the role of the government and not the insurer, whose role is to provide affordable insurance, as stated by the CCS stakeholders interviewed. The agricultural insurance in Spain is also similar, in that it does not have a strong focus on promoting risk reduction by policyholders. Rather, these markets as a whole rely on government-provided risk reduction to maintain flood risk at an insurable level, such as a 1 in 75 year protection standard in the UK.

For agricultural insurance, we see that the best practice countries of Austria and Spain both employ bonus-malus systems which reward policyholders with a reduction in premiums if insurance claims lie below a certain level and increase premiums if otherwise. Policyholders are indirectly rewarded for undertaking actions that lower long-run extreme weather event vulnerability. This approach allows for a certain degree of information asymmetry between the insurer and the policyholder, as the insurer cannot observe and evaluate all the actions that a policyholder may take, but it can monitor claims as a proxy measurement. The Spanish case goes beyond the bonus-malus system by offering two additional risk-reduction strategies. First, to be insured a farmer must meet certain pre-set conditions regarding vulnerability to extreme events. Second, premium discounts are given in return for employing risk-reduction measures.

Overall, the case study countries produce low scores in the MCA due to their reliance on incentives that provide indirect risk management signals such as awareness campaigns and deductibles. Awareness campaigns attempt to alter risk perceptions to improve the perceived benefits of implementing risk-reduction measures. For example, the German stakeholders engaged with during the project provided anecdotal evidence that their efforts in collaboration with local and federal governments in Germany through the *Elementarschadenskampagne* (elemental damage campaign) risk communication campaign helped to increase the flood insurance penetration rate across Germany.

3.1.2. Implications for policies to improve resilience

As noted in the previous section, the best practice cases identified by the MCA do not necessarily result in a systematic improvement for the resistance resilience pillar. For instance, the information campaign that the German stakeholders said was effective has its limitations as, to the best of our knowledge; a systematic evaluation of the campaign has not been conducted. Moreover, there are limitations to relying on awareness campaigns, because constant campaigns are required in order to prevent risk perceptions from declining due to availability biases (i.e. a tendency for experiences to be perceived as less important the longer ago they occurred). Moreover, the limited academic literature also displays mixed results. For instance, Osberghaus (2017) finds that in Germany more educated households are more responsive to flood risk adaption. While not directly related to risk communication, this finding indicates that there can be a positive response to education campaigns. Maidl and Buchecker (2015) used survey data to evaluate the success of a risk communication campaign aimed at increasing flood preparedness. Their study found that the campaign was able to increase preparedness intentions, though its success was based upon having a positive view of the material received as well as possessing a sufficient level of awareness. Rollason, Bracken, Hardy, and Large

(2018) discuss a risk communication strategy in the UK, finding that these communication strategies should be tailored to local contexts in order to be successful. Therefore, while anecdotal evidence suggests that the large-scale risk communication campaign in Germany has been successful, there should be more efforts to evaluate the success of such campaigns.

Secondly, we see that the link between insurance and the resistance resilience pillar is hampered by the reliance on deductibles, particularly when they are small. Deductibles alone are unlikely to incentivise proactive risk reduction because policyholders need to correctly understand the amount of their level of extreme weather risk that is not covered by the insurer before this can act as an incentive to lower risk. Empirical research has indicated that only large deductibles can act as an incentive for limiting natural disaster risk (Hudson, Botzen, Czajkowski, & Kreibich, 2017) and that more tangible incentives may be more successful. For example, Mol, Botzen, and Blasch (2018) find, via experiments, that premium discounts appear successful at encouraging policyholders to invest in flood damage mitigation measures. On the other hand, it has been noted in Australia that premiums do not provide a strong incentive to boost resistance due to the non-transparency of risks and premiums (Mcaneney et al., 2016).

Nevertheless, the best practice examples do offer directions for how other insurance markets could increase the resistance pillar of resilience. The linkages between insurance and the resistance pillar can be improved by developing a series of more direct mechanisms to incentivise risk reduction. Such mechanisms can be layered to offer a different range of possible incentives, like information provision about risk and effective risk-reduction measures with premium discounts for implementing these measures. This was found to be the case in Austria and Spain for crop insurance, as a layering of resistance boosting mechanisms is an aspect of the best practice examples that can be used across both private property insurance and crop insurance. The layered incentive arrangement allows the insurer to signal which measures are known to be effective. Moreover, this may be achieved while promoting decentralised adaptation through premium discounts, which should only promote adaptation when adaptation is cost-effective. Secondly, bonus-malus arrangements may help overcome elements of asymmetric information.

Therefore, it is beneficial to develop a wider portfolio of risk incentivisation mechanisms in collaboration with a wider range of stakeholders in order to create a suitable environment. For instance, the Department for Environment, Food and Rural Affairs in the UK commissioned reports looking into what could be suitable low-cost package of measures and strategies that homeowners can use to limit flood risk (Lamond, Rose, Bhattacharya-Mis, & Joseph, 2018). This was later further refined through a series of reports for the UK insurance pool Flood Re looking into how their use can be incentivised (Flood Re, 2018) and which ones are likely to be cost-effective (Lamond et al., 2018). Due to Flood Re often being criticised for a lack of focus on risk reduction (e.g. Surminski, 2018; Surminski & Eldridge, 2015), these actions by Flood Re indicate a promising direction for strengthening the resistance pillar of insurance in the UK. Moreover, in order to be successful at promoting the resistance pillar of resilience these insurance-based incentives should be part of a wider enabling environment and partnership arrangement, as insurance by itself is insufficient (Linnerooth-Bayer et al., 2019; Surminski & Thieken, 2017). However, there tends to be slow movement in bringing stakeholders together on this topic (see section 3.4.2), even though multi-sector engagement is required (Kunreuther, 2015).

Additionally, a concern for the growth of flood risk involves the use of land-use planning and how governments do not correctly consider how changes in land-use may impact flood risk (see Mcaneney et al. (2016) or (OECD, 2016)). However, while land-use planning is not predominantly within the insurance sphere of influence, the identified best practice countries highlight cases where land-use can still be influenced. For instance, in the UK, Flood Re prohibits the sale of insurance for buildings in high flood risk zones constructed after 2008. This is in order to provide an incentive to limit development in high-risk floodplains in the coming years, due to the common insurance requirements compelled by UK mortgage conditions. Therefore, while insurance may not be able to directly influence land-use planning, insurers may be able to steer land-use in a more resilient direction through their insurance requirements.

3.2. Recovery

3.2.1. Outcome of the MCA

The European insurance markets perform best when considered primarily as mechanisms focusing upon recovery, as seen by the highest MCA scores for Weighting Scheme 1. The best practice countries achieve high MCA scores because of high penetration rates, comprehensive multi-hazard coverage, and mechanisms in place to support large loss events. This combination of factors allows the insurance markets of France, Spain, and the UK to be in, overall, a good position to provide clear and secure compensation for private property insurance after a flood event. Spain, Austria, and Sweden provide similar arrangements regarding crop insurance. This is due to their use of multi-hazard crop insurance and a compulsion to insure all farmland rather than selecting only higher risk parcels.

In comparing the features of the three best practice cases in the private property and crop insurance sectors, we see that in both market sectors the best practice cases bundle several extreme weather risks into a single policy. However, it is unclear in these cases to what degree each extreme weather event contributes towards the insurance premium. In combing several risks together, a greater and more diversified pool of policyholders can be created. Moreover, these countries tended to have a formal or informal mandate to buy insurance. For instance, extreme weather insurance is a compulsory extension of general private property insurance in Spain and France. These two features taken together increases the ability of the extreme weather insurance markets to display the mutuality and solidarity needed to function (Linnerooth-Bayer et al., 2019).

While mandating multi-extreme weather event coverage provides wider access to the recovery mechanism of insurance, this in turn must be balanced with concerns about affordability. This trade-off is especially important if there is a wider movement towards risk-based premiums to provide stronger risk reduction incentives or if voluntary purchase options are retracted. The Spanish and French cases aim to limit unaffordability by directly linking the natural hazard premium to a fixed percentage of coverage brought. The Danish and UK cases are less clear on this topic. Although it is difficult to distinguish which part of the premium reflects which risk, it is clear that there is a large degree of cross-subsidisation between higher and lower risk households. However, a key observation regarding affordability is that the main driver, on average, of insurance unaffordability is the buying power of households rather than the amount of the insurance premiums. For example, the average rate of unaffordability in Bulgaria is estimated to be 23%, while in Romania the rate is 26%, but the annual average premiums are only €90 and €30, respectively (Insurance Europe, 2015).

Finally, in order to support insurers in the wake of large events, the best practice countries have mechanisms in place that help to maintain solvency. France provides reinsurance for extreme weather events through a governmentally provided reinsurance facility, which in turn has an unlimited state guarantee. In the UK, however, private reinsurers have created the Flood Re pool to share losses between participating insurers, which in turn has quasi-governmental powers to raise levies to provide a suitable capital base.

3.2.1. Implications for policies to improve resilience

There are several implications of the conclusions drawn regarding how insurance best practice interacts with the recovery pillar of resilience. In principle, the provision of insurance greatly supports the recovery pillar, because policyholders are no longer solely reliant on their own resources (i.e. self-insurance) to get back on their feet after an extreme weather event. However, for this enhancement to be truly embraced and actively improve the recovery pillar of resilience, insurance coverage must be widespread.

The best practice countries were able to achieve this in both the private property and crop insurance markets due to a commonly observed effective compulsion to buy a comprehensive bundle of extreme weather event insurance. Moreover, in the case of Spain the potential for access to government compensation funds is only available for those who have purchased sufficient agricultural insurance coverage. This collection of features allows relatively rare extreme weather events, such as floods, to be pooled with more common extreme weather events such as hailstorms, as well as uncorrelated risks, such as house fires, which are more tangibly important to the policyholder. This combination of perils allows for improved risk diversification by increasing the range of high and low risk policyholders. Moreover, broad coverage of a variety of risks gives policyholders a high degree of certainty in receiving compensation after an extreme weather event occurs. These structured systems promote a reliance on formally developed and provided insurance mechanisms rather than ad hoc government support. Furthermore, promoting a reliance on formal and developed mechanisms promotes a higher level of development of the recovery resilience pillar because there are predictable expectations rather than actions based on changing public concerns and pressures. This is because mandated and comprehensive insurance coverage is more efficient than a system of ex post public disaster programmes (Kunreuther, 2006).

Therefore, in order to promote increased rates of coverage in the non-best practice regions of Europe, there could be an increased focus on mandates to buy a complete bundle of extreme weather event insurance. The development of such a mandate is not sufficient without enforcement or sufficient incentives for buying insurance. Therefore, linking multi-hazard insurance with products that a consumer sees as more important can increase resilience by increasing the opportunity cost of not buying comprehensive coverage. Private property insurance has a well-developed mechanism that links extreme weather insurance with mortgage requirements or fire insurance. While a similar observation was not made for farmers, it was found in the case of Spain that holding sufficient insurance coverage was a requirement for being able to receive potential compensation from the government. Therefore, we argue that this approach can be extended so that insurance coverage is tied to access to wider agricultural subsidies (e.g. the Common Agricultural

Policy) to increase the opportunity cost of not being insured. These suggestions for moving insurance forwards are common place in the countries identified among the selected cases as exemplars of best practice (although not in the worst performing countries) as well as several other countries in the world. Additionally, the relative salience of the risks of these events can be improved by making them more tangible, for example, by connecting the occurrence of flooding to the lifetime of a mortgage or a resident's tenure in their property. This, in turn, could lead to a stronger demand for multi-year insurance policies, which could create a more stable extreme weather insurance demand for insurers, as argued by Michel-Kerjan, Lemoyne De Forges, and Kunreuther (2012).

The above mechanisms can help boost the recovery pillar resilience and decrease the difficultly for insurers of insuring localised disasters by increasing currently limited demand. However, in doing so care must be taken to maintain a degree of clarity about how the premium is structured and connected to the types of insured extreme weather risks, to avoid the problems encountered in Australia regarding risk reduction (Mcaneney et al., 2016).

Not only is the clarity of the premium important, but also, in terms of the social justice implications when coverage is mandated, its total size. This is because if premiums are considered unaffordable then the purchaser is considered to be overly burdened by the purchase (National Research Council, 2015). Unaffordability in the case study countries is driven, in most cases, by the purchasing powers of households rather than the premiums themselves. However, if premiums are linked to risk, then premiums can become rapidly unaffordable in high risk areas, unless risk-reduction measures are taken either by the state, community, or individual. A high rate of unaffordability places additional burdens on policyholders, which limits their resilience potential. Therefore, as the results of the MCA indicate that unaffordability cannot be fully corrected from within the insurance market itself, mechanisms external to the market are required to support policyholders who would face unaffordable insurance premiums.

A commonly proposed method for addressing this problem is means-tested vouchers for enabling low-income households in high risk areas to purchase comprehensive insurance (Kousky & Kunreuther, 2014). Such vouchers can address equity concerns by removing the high premium burden placed on households while potentially allowing for insurance to have a stronger link to risk reduction. This is because vouchers allow unaffordability to be corrected from outside of the insurance market. The same can also hold for property-level adaptation measures in order to render them more affordable and increase uptake (Montgomery & Kunreuther, 2018). The voucher mechanism, or similar ones such as tax credits, allow for insurance to improve two pillars of resilience before and after extreme weather events. However, such mechanisms for purchasing insurance should also be time-limited (Kousky & Kunreuther, 2014) in order to avoid an indirect subsidy for lower-income households to locate in disaster prone areas in order to become eligible for the voucher. Including these social justice concerns can help to limit social inequalities that can occur if only certain segments of society can successfully adapt to extreme weather events.

However, a relevant concern is that in the case of a large event rapid insurance pay-outs may not be possible due to the assessment and claims process. For instance, this was a problematic experience after the Christchurch earthquake where, due to the magnitude of the event, payments were staggered over many years (King, Middleton, Brown,

Johnston, & Johal, 2014; Potter, Becker, Johnston, & Rossiter, 2015), as well as after Hurricane Katrina (Corey & Deitch, 2011; Green, Bates, & Smyth, 2007). This process can be further complicated by the presence of both private insurers and public compensation mechanisms. Therefore, it is important to correctly layer an insurance market so that those affected by a disaster have a single point of call for claiming assistance. This line of thought could also support the development of PPPs where the government acts as a formal reinsurer rather than offering direct compensation to people and thereby acting as a competitor to primary insurers. Creating such an enabling environment for facilitating fast payments is important, as Poontirakul, Brown, Seville, Vargo, and Noy (2017) notes that adequate and prompt payments promote recovery, while slow payment processes may be no better than receiving no compensation payments at all.

3.3. Adaptive capacity

3.3.1. Outcome of the MCA

There are several ways in which insurance can interact with the concept of adaptive capacity. The first is the ability of the insurance industry to inform policyholders through awareness-increasing activities, as discussed in Section 3.1. The second is through the ability to remain a viable adaptation mechanism moving forwards in light of increasing natural hazard losses. The best practice cases identified have extensive collaboration across stakeholders, which have allowed the insurance arrangements to adapt to changing situations in a broadly acceptable manner. For example, the Danish Storm Council is appointed by the Danish Minister for Business and Growth and consists of an independent chair and eight other members, in order to bring a holistic and collaborative understanding of risk management topics. These include stakeholders from the insurance industry, private citizens, municipalities and ministries. The Spanish approach to crop insurance is based around the Agroseguro entity that handles the entire insurance process, with the objective of managing agricultural risk as a whole for its stakeholders.

A challenge for transferring these best practices to other countries is that, for the most part, the institutions are long established. The Austrian Hail Insurance Company was founded in 1946, Agroseguro was founded in 1978, and the CCS can trace its origins to 1941. The most recently formed organisation is Flood Re in the UK. The introduction of Flood Re highlights several potential problems surrounding different levels of willingness to participate between insurers and the government as well as potential legal challenges, which is described in more detail by Mysiak and Pérez-Blanco (2016) and Surminski (2018). Taken together, the long tradition of a central body helps create a suitable enabling environment for insurance markets to operate in. However, such bodies can be very difficult to create without a suitable catalysing event, such as CATNAT and Flood Re being founded as a response to serious events. A discussion of the role of catalysing events is presented by Birkmann et al. (2010).

3.3.2. Implications for policies to improve resilience

While the resistance-improving aspect of the insurance best practice cases is the weakest aspect that can be formally measured through the MCA, the adaptive capacity elements are also weak. For instance, there is little evidence regarding the success at increasing resilience of risk communication campaigns. Moreover, the insurance mechanisms studied tend to be rather static in terms of their overall structure. This hints that the interaction between insurance and adaptive capacity is underutilised. This is the case despite the collaborative structures in place that provide insurers with suitable support in case of large disaster events, which otherwise may threaten their solvency. Insolvent insurers would not be able to provide compensation to those affected. For example, in Spain the CCS has an equalisation fund financed by retained premiums, and on top of this fund the CCS can buy private reinsurance coverage and has access to a state guarantee. Similarly, Flood Re in the UK acts as an insurance pool financed by a surcharge on all insurance policies sold in the UK, with the ability to impose a second surcharge in case Flood Re's resources prove insufficient. These measures improve adaptive capacity by maintaining insurer solvency regarding high risk households.

Furthermore, the degree of collaboration between the insurance sector stakeholders and the government can be tailored with respect to formality and extent of the partnership. For instance, in France the public and private sector have a long-standing cooperation in the French Association for Disaster Risk Reduction. Another example is the Danish Storm Council. However, regardless of the degree of formality, there is a stated focus. Therefore, a suitable role for new bodies can in promoting risk reduction. This is because an external body dedicated to promoting and developing risk reduction strategies integrated across all relevant stakeholders can facilitate a minimum level of risk management and insurance viability. This could be achieved by adding a surcharge to insurance premiums that channels the revenue into a fund or funds for constructing protection measures, for general adaptation measures, or to subsidise more individual property-level measures. This fund could be a coordination entity whereby insurers, government agencies, etc. are involved in a not-for-profit manner.

In developing a coordination entity, the specific knowledge and expertise of each stakeholder can be leveraged to strengthen current weaknesses that a single stakeholder cannot surmount. For example, insurers can use their expertise to identify which adaptation measures lower risk, and a third-party organisation can provide certificates to those who employ these measures to signal that these measures have been correctly employed. This, in turn, allows the policyholder to obtain discounts on their insurance premiums or reimbursement for the measure's cost. This approach is similar to the elevation certificates offered for the National Flood Insurance Program in the USA, or the Texas Windstorm Insurance Association's WPI-18 certificates, as mentioned by Mcaneney et al. (2016). A single management body can organise and facilitate such stakeholder collaborations. Moreover, such a body brings together a range of experiences and capacities that can be used to develop the socially inclusive and useful risk communication campaigns required for the message to be acted upon.

4. Conclusion

Extreme weather events place a large burden on society due to their potentially disastrous consequences. Moreover, due to the combination of socio-economic development and climate change there is a growing threat from extreme weather events. Therefore, societal resilience against extreme weather events should be promoted and, if well organised and regulated, insurance is, potentially, a transformative mechanism for resilience.

The current role that extreme weather event insurance plays varies strongly across Europe, with a great deal of heterogeneity in provision and overall outcomes across countries and sectors. A comparison of this diversity can identify what works well and what does not, with the aim of deriving recommendations for improving the performance of insurance markets to enhance extreme weather resilience. To conduct our comparative study, we used an extensive inventory of insurance markets covering 12 countries for private property and agricultural insurance across a range of extreme weather events. The comparative analysis identified that, for private property insurance, Denmark, France, Spain, and the UK represent exemplars of best practice. For the agriculture sector best practice exemplars were Austria, Spain, and Sweden.

Based on this analysis, from the policymaker perspective, we suggest a series of ways in which extreme weather event insurance can increase disaster resilience. For private property insurance, we suggest introducing a requirement to buy extreme weather event insurance along with general homeowners' insurance. Moreover, support for low income households through means-tested vouchers could be provided. For the agricultural sector, we propose moving towards comprehensive crop yield insurance by requiring farmers to buy a sufficiently comprehensive insurance product to be eligible for the general agricultural subsidies farmers receive. In both cases, a nationally representative body can coordinate the various stakeholders into acting in concert. This body could be financed by a premium surcharge which is then used to directly co-finance the employment of cost-effective risk-reduction measures. Moreover, this public-private collaboration could aim to strengthen the link between insurance and risk reduction through a combination of measures, including information provision about risk and mitigation measures, financial incentives like premium discounts and subsidies, and building codes and zoning regulations. In the long run, improved risk reduction will result in lower premiums. Additionally, this structure could help to support the financing of extreme losses to maintain the solvency of the insurance industry. This suggested series of policy recommendations for boosting resilience is similar to features of the reforms proposed for the National Flood Insurance Program in the United States by Michel-Kerjan and Kunreuther (2011). This shows the overall generalisability of these results and policy directions at the global level regarding how insurance should move forwards to enhance resilience.

Note

1. Please note that earthquakes or other seismic risks, while present in Europe, are not relevant to this study as it focuses on extreme weather events. Future research can address this gap in the literature.

Disclosure statement

No potential conflict of interest was reported by the authors.

Funding

This work was supported by Climate Adaption Unit, DG CLIMA, European Comission. Botzen has recieved support from the Netherlands Organisation for Scientific Research (NWO) VIDI Grant (452.14.005) and the European Union's Horizon 2020 research and innovation programme under grant agreement No 776479.

References

Aerts, J. C. J. H., Botzen, W. J. W., De Moel, H., & Bowman, M. (2013). Cost estimates for flood resilience and protection strategies in New York City. *Annals of the New York Academy of Sciences, 1294*, 1–104.

Birkmann, J., Buckle, P., Jaeger, J., Pelling, M., Setiadi, N., Garschagen, M., … Kropp, J. (2010). Extreme events and disasters: A window of opportunity for change? Analysis of organizational, institutional and political changes, formal and informal responses after mega-disasters. *Natural Hazards, 55*, 637–655.

Botzen, W. J. W. (2013). *Managing extreme climate change risks through insurance*. New York: Cambridge University Press.

Burby, R. (2001). Flood insurance and floodplain management: The US experience. *Global Environmental Change Part B: Environmental Hazards, 3*, 111–122.

Charpentier, A. (2008). Insurability of climate risks. *The Geneva Papers, 33*, 91–109.

Chatterton, J., Clarke, C., Daly, E., Dawks, S., Elding, C., Fenn, T., … Salado, R. (2016). *The costs and impacts of the winter 2013 to 2014 floods*. Bristol: Environment Agency.

Corey, C. M., & Deitch, E. A. (2011). Factors affecting business recovery immediately after Hurricane Katrina. *Journal of Contingencies and Crisis Management, 19*, 169–181.

Cremades, R., Surminski, S., Máñez Costa, M., Hudson, P., Shrivastava, P., & Gascoigne, J. (2018). Using the adaptive cycle in climate-risk insurance to design resilient futures. *Nature Climate Change, 8*, 4–7.

European Comission. (2017). Insurance of weather and climate-related disaster risk: Inventory and analysis of mechanisms to support damage prevention in the EU. In X. Le Den, M. Persson, A. Benoist, P. Hudson, M. De Ruiter, L. De Ruig, O. Kuik, & W. Botzen (Eds.). Luxembourg: Publications Office of the European Union.

Flood Re. (2018). *Incentivising household action on flooding and options for using incentives to increase the take up of flood resilience and resistance measures*. Author.

The Geneva Association. (2018). *Climate change and the insurance industry: Taking action as risk managers and investors perspectives from C-level executives in the insurance industry*. Zurich: Author.

Golnaraghi, M., Surminski, S., & Schanz, K. (2017). *An intergrated appraoch to managing extreme events and cliamte risks*. Zurich: The Geneva Assocaition.

Green, R., Bates, L. K., & Smyth, A. (2007). Impediments to recovery in New Orleans' upper and lower ninth ward: One year after Hurricane Katrina. *Disasters, 31*, 311–335.

Hochrainer-Stigler, S., & Lorant, A. (2018). Evaluating partnerships to enhance disaster risk management using multi-criteria analysis: An application at the Pan-European level. *Environmental Management, 61*, 24–33.

Hudson, P., Botzen, W. J. W., Czajkowski, J., & Kreibich, H. (2017). Moral hazard in natural disaster insurance markets: Empirical evidence from Germany and the United States. *Land Economics, 93*, 179–208.

Hudson, P., Botzen, W. J. W., Feyen, L., & Aerts, J. C. J. H. (2016). Incentivising flood risk adaptation through risk based insurance premiums: Trade-offs between affordability and risk reduction. *Ecological Economics, 125*, 1–13.

Insurance Europe. (2015). *European insurance - key facts*. Brussels: Author.

King, A., Middleton, D., Brown, C., Johnston, D., & Johal, S. (2014). Insurance: Its role in recovery from the 2010–2011 Canterbury earthquake sequence. *Earthquake Spectra, 30*, 475–491.

Kousky, C., & Kunreuther, H. (2014). Addressing affordability in the national flood insurance program. *Journal of Extreme Events, 1*, 1450001.

Kreibich, H., Thieken, A. H., Petrow, T., Müller, M., & Merz, B. (2005). Flood loss reduction of private households due to building precautionary measures- lessons learned from the Elbe flood in August 2002. *Natural Hazards and Earth System Science, 5*, 117–126.

Kunreuther, H. (2006). Disaster mitigation and insurance: Learning from Katrina. *The ANNALS of the American Academy of Political and Social Science, 604*, 208–227.

Kunreuther, H. (2015). The role of insurance in reducing losses from extreme events: The need for public-private partnerships. *The Geneva Papers on Risk and Insurance Issues and Practice, 40*, 741–762.

Kunreuther, H. (2017). *Encouraging adaptation to flood risk: The role of the national flood insurance program* (Wharton Working Papers). Philadelphia: Wharton, University of Pennsylvania.

Kunreuther, H., & Michel-Kerjan, E. (2009). *At war with the weather: Managing large scale risks in a new era of catastrophes*. Cambridge, MA: MIT Press.

Kunreuther, H., & Michel-Kerjan, E. (2013). Managing catastrophic risks through redesigned insurance: Challenges and opportunities. In G. Dionne (Ed.), *Handbook of insurance* (pp. 517–546). New York, NY: Springer New York.

Lamond, J., Rose, C., Bhattacharya-Mis, N., & Joseph, R. (2018). *Evidence review for property flood resilience phase 2 report*. Bristol: Flood Re and UWE Bristol.

Linnerooth-Bayer, J., Surminski, S., Bouwer, L. M., Noy, I., & Mechler, R. (2019). Insurance as a response to loss and damage? In R. Mechler, L. M. Bouwer, T. Schinko, S. Surminski, & J. Linnerooth-Bayer (Eds.), *Loss and damage from climate change: Concepts, methods and policy options* (pp. 483–512). Cham: Springer International Publishing.

Maidl, E., & Buchecker, M. (2015). Raising risk preparedness by flood risk communication. *Natural Hazards and Earth System Sciences, 15*, 1577–1595.

Mas-Colell, A. (1995). *Microeconomic theory*. New York: Oxford University Press.

Mcaneney, J., Mcaneney, D., Musulin, R., Walker, G., & Crompton, R. (2016). Government-sponsored natural disaster insurance pools: A view from down-under. *International Journal of Disaster Risk Reduction, 15*, 1–9.

Michel-Kerjan, E., & Kunreuther, H. (2011). Redesigning flood insurance. *Science, 333*, 408–409.

Michel-Kerjan, E., Lemoyne De Forges, S., & Kunreuther, H. (2012). Policy tenure under the U.S. National Flood Insurance Program (NFIP). *Risk Analysis, 32*, 644–658.

Mol, J. M., Botzen, W. J. W., & Blasch, J. E. (2018). Behavioral motivations for self-insurance under different disaster risk insurance schemes. *Journal of Economic Behavior & Organization*. https://doi.org/10.1016/j.jebo.2018.12.007

Montgomery, M., & Kunreuther, H. (2018). Pricing storm surge risks in Florida: Implications for determining flood insurance premiums and evaluating mitigation measures. *Risk Analysis, 38*, 2275–2299.

Munich Re. (2017). *Natural catastrophes 2016 Analyses, assessments, position. TOPICS GEO*. Munich, Germany: Author.

Mysiak, J., & Pérez-Blanco, C. D. (2016). Partnerships for disaster risk insurance in the EU. *Natural Hazards and Earth System Sciences, 16*, 2403–2419.

Mysiak, J., Surminski, S., Thieken, A., Mechler, R., & Aerts, J. (2016). Brief communication: Sendai framework for disaster risk reduction – success or warning sign for Paris? *Natural Hazards and Earth Systems Science, 16*, 2189–2193.

National Research Council. (2015). *Affordability of national flood insurance program premiums: Report 1*. Washington, DC: National Research Council.

OECD. (2016). *Financial management of flood risk*. Paris: OECD.

Osberghaus, D. (2017). The effect of flood experience on household mitigation—evidence from longitudinal and insurance data. *Global Environmental Change, 43*, 126–136.

Poontirakul, P., Brown, C., Seville, E., Vargo, J., & Noy, I. (2017). Insurance as a double-edged sword: Quantitative evidence from the 2011 Christchurch earthquake. *The Geneva Papers Risk and Insurance - Issues and Practice, 42*, 609–632.

Potter, S. H., Becker, J. S., Johnston, D. M., & Rossiter, K. P. (2015). An overview of the impacts of the 2010–2011 Canterbury earthquakes. *International Journal of Disaster Risk Reduction, 14*, 6–14.

Poussin, J. K., Botzen, W. J. W., & Aerts, J. C. J. H. (2013). Stimulating flood damage mitigation through insurance: An assessment of the French CatNat system. *Environmental Hazards, 12*, 258–277.

Rollason, E., Bracken, L. J., Hardy, R. J., & Large, A. R. G. (2018). Rethinking flood risk communication. *Natural Hazards, 92*, 1665–1686.

Schäfer, L., Warner, K., & Kreft, S. (2019). Exploring and managing adaptation frontiers with climate risk insurance. In R. Mechler, L. M. Bouwer, T. Schinko, S. Surminski, & J. Linnerooth-Bayer (Eds.), *Loss and damage from climate change: Concepts, methods and policy options* (pp. 317–341). Cham: Springer International Publishing.

Surminski, S. (2018). Fit for Purpose and Fit for the Future? An Evaluation of the UK's New Flood Reinsurance Pool. *Risk Management and Insurance Review, 21*, 33–72. doi:10.1111/rmir.12093

Surminski, S., Aerts, J. C. J. H., Botzen, W. J. W., Hudson, P., Mysiak, J., & Pérez-Blanco, C. D. (2015). Reflection on the current debate on how to link flood insurance and disaster risk reduction in the European Union. *Natural Hazards, 79*, 1451–1479.

Surminski, S., & Eldridge, J. (2015). Flood insurance in England: An assessment of the current and newly proposed insurance scheme in the context of rising flood risk. *Journal of Flood Risk Management, 10*, 415–435.

Surminski, S., & Oramas-Dorta, D. (2014). Flood insurance schemes and climate adaptation in developing countries. *International Journal of Disaster Risk Reduction, 7*, 154–164.

Surminski, S., & Thieken, A. (2017). Promoting flood risk reduction: The role of insurance in Germany and England. *Earth's Future, 5*, 979–1001.

Thieken, A. H., Mariani, S., Longfield, S., & Vanneuville, W. (2014). Preface: Flood resilient communities–managing the consequences of flooding. *Natural Hazards and Earth System Science, 14*, 33–39.

Unterberger, C., Hudson, P., Botzen, W. J. W., Schroeer, K., & Steininger, K. W. (2019). Future public sector flood risk and risk sharing arrangements: An assessment for Austria. *Ecological Economics, 156*, 153–163.

Disaster, relocation, and resilience: recovery and adaptation of Karamemedesane in Lily Tribal Community after Typhoon Morakot, Taiwan

Sasala Taiban, Hui-Nien Lin and Chun-Chieh Ko

ABSTRACT

After Typhoon Morakot struck Taiwan in the summer of 2009, government officials relocated the indigenous village communities of Kucapungane, Adiri, Karamemedesane, Kinulane, Dawadawan and Tikuvulu into sub-montane, permanent housing. Because villagers were accustomed to living in mountainous areas, they encountered many challenges while adapting their lifestyle and culture into a new setting. During the relocation process, government and post-disaster relief agencies disregarded, oversimplified, and concealed social vulnerability. Can indigenous communities recover from typhoon damage and continue to pass down their culture? Using in-depth interviews and participant observation, this research examined how Karamemedesane villagers organised and reconstructed themselves using their land for farming practices, culture, rituals, and livelihoods following the government-forced, community migration. The source of resilience for Karamemedesane turned out to be the cultivation of red quinoa, a traditional food crop. Villagers rediscovered the cultural value of food through small changes in farming practices and knowledge, social network and social learning, leadership, and innovation-aided recovery that resulted in establishing the Academy of Special Rukai Crops. Results suggested that post-disaster policies for indigenous communities should be land-based and culturally relevant to promote transformability.

Introduction

During August 2009, Typhoon Morakot produced more than 2000 mm of rainfall in two days and caused catastrophic damage to the central and southern parts of Taiwan. In the aftermath, 673 people died; 26 went missing; and 1766 houses were destroyed.

To facilitate recovery and reconstruction after the disaster, the Legislative Yuan, Republic of China (ROC), approved the third reading of the Special Act of Reconstruction after Typhoon Morakot (SARTM) on 27 August 2009. According to Article 20 of the SARTM,

> The central government or governments of special municipalities, counties, or cities may, by reaching a consensus with the original occupant(s), delimit specific disaster region(s) as

involving safety hazards or unlawful constructions and impose residency restrictions or forced relocation of villagers within the specified time limit. The government(s) shall provide proper resources for relocation in fulfilment of the abovementioned requirements.

Delimiting the disaster area was controversial for affected indigenous people and resulted in severe obstacles for them to overcome after relocation (Huang, 2018a; 2018b; Taiban, 2012; 2013). Nevertheless after Typhoon Morakot, Lily Tribal Community (LTC) was selected as one of the sites for permanent housing for Rukai people from the Adiri, Kinulane, Karamemedesane, and Kudrengere villages from Wutai Township and Paiwan people of the Dawadawan and Tikuvulu villages from Sandimen Township. The community name was Lily (i.e. 'Baihe' in Chinese) because this flower is a cultural symbol shared by Rukai and Paiwan, the two ethnic groups of LTC residents. Lilies also have six petals, representing the six LTC communities (i.e. Adiri, Kinulane, Karamemedesane, Kudrengere, Dawadawan, and Tikuvulu), thus symbolising unity.

Traditionally, Rukai people divide their living areas into three zones based on elevation and average temperature. Areas with an elevation of ≥1500 m belong to *Rekai*, a zone with dry and cold air; those with an elevation of 500–1000 m belong to *Paralibicane*, a zone with a temperate climate; and those with an elevation of <500 m comprise *Labelabe*, a zone with hot and humid air. Rukai call themselves 'people who live in cold and dry areas' and regard *Labelabe* as having unsuitable living conditions due to excessive heat and humidity (Taiban, 2008). However, the programme for large-scale migration after Morakot required villagers from Adiri, Kinulane, Karamemedesane, and Ira to move to LTC, which is less than 100 m in elevation (and therefore similar to the *Labelabe* zone).

The Karamemedesane community, originating from a branch of the western Rukai group (or the Ailiao group), was the focus of this research. After Typhoon Morakot, government officials sub-divided the Karamemedesane village on the mountainside, declaring safe and unsafe zones for human habitation. Residents of the first, second, and third neighbourhoods (59 unsafe households) were relocated to LTC, whereas villagers in the fourth and fifth neighbourhoods (54 safe households) could remain on-site. The case of Karamemedesane can therefore contribute to research on environmentally forced migrations (Renaud, Bogardi, Dun, & Warner, 2007; Renaud, Dun, Warner, & Bogardi, 2011). Sub-dividing Karamemedesane village by SARTM raises an important question. How does community resilience play a role in the recovery process between villagers who were split between two sites (LTC and the original village)? Nearly a decade after relocation how did the Karamemedesane villagers adjust, adopt and possibly unite through the post-disaster recovery?

Concepts from the research literature

Environmental disasters and social vulnerability

Human focused discussions of environmental disasters have deepened and broadened due to the prevalence of extreme weather events in the modern world (Blaikie, Cannon, Davis, & Wisner, 1994; Hewitt, 1983; Oliver-Smith & Hoffman, 1999). Although scientists have studied disasters since the 1990s, there is little agreement on terms and terminology. Oliver-Smith and Hoffman (1999) defined 'hazard' as a type of power, technology, or condition that can damage a society, infrastructure, or the environment, whereas 'disaster' was defined as environmental factors that can be destructive and may lead to social vulnerability.

Blaikie et al. (1994) proposed more systemic definitions for disasters and hazards. Disasters arise from three mutually influential elements: hazard, risk, and vulnerability. By contrast, hazards are physical factors that underline disasters and are predictable through statistical research. Risk refers to the sum of vulnerability of an area that results from complex and known environmental hazards. Vulnerability relates to the extent to which an individual or group can predict, deal with, resist, and recover from the effects of an environmental hazard. Blaikie asserted that the causes of disasters are rooted in a given cultural context, historically and structurally.

From vulnerability to resilience: an innovative perspective for disaster research

Vulnerability focuses on the effects of various social and economic conditions on groups of people in the context of disaster risks. Such analyses have indicated that when disasters occur, differences in casualties and economic loss among groups can be traced to ethnic, social status, gender, and regional inequities. Multiple perspectives of vulnerability may be useful to explore the historical and social pre-disaster factors, leading to different levels of effects.

After a disaster occurs, societies and individuals tend to vary in their reactions; therefore, vulnerability varies. For example, when coping with the effects of Typhoon Morakot, some people were able to support one another in joint rescues of trapped families, friends, or neighbours, resulting in a lower number of causalities, as compared to other affected villages. A number of villages received assistance quickly by utilising social networking sites, leading to early relocation and reconstruction efforts. Several communities discovered alternative approaches to development. In some cases, these alternative approaches led to greater development than the villagers had experienced in their pre-disaster condition. However, other villages declined assistance following the disaster. Community resilience of affected groups is crucial for rapid recovery from disasters and adaptation to altered conditions. Some authors have noted that the main objective of disaster research has shifted from casualties and loss, to community recovery and post-disaster growth (Paton & Johnston, 2001).

Adaptation is anthropological term used to describe successful or 'functional' interactions of human cultures in localised environments as part of a long-term evolutionary process (Cohen, 1974; Nelson, West, & Finan, 2009). In a shorter time span, Nelson, Adger, and Brown (2007) defined adaptation as decision-making processes and the set of actions undertaken to deal with perturbations of social-ecological systems without any significant changes while preserving future options. Over the past few decades, there has been an increase in the frequency and severity of disasters such as droughts, flooding and cyclones. The impacts on development, poverty and vulnerability have led to calls for improving resilience – meaning enhancing the capacity of households, communities and countries to cope with, and adapt to shocks and stresses associated with environmental hazards.

The concept of resilience comes from ecology, meaning the ability to cope with stress without changing a community's structure and function, while increasing its ability to withstand future pressures (Brown & Kulig, 1996; Holling, 1973). In other words, resilience refers to the ability to predict risk, limit impact, and recover quickly by survival, adaptability, evolution, and growth in the face of turbulent change (Pelling, 2003a).

In the realm of the social sciences, discussions of resilience originated from anthropological studies conducted in the 1980s; specifically, research of settlement development using ethnology. In a study comparing post-flood responses within Guyana, South America, Pelling (2003b) indicated that traditional fishing villages exhibited greater resilience against floods than did modern cities with professional planners. Villagers had developed effective flood control methods after many past failures, and had adapted their lifestyles accordingly. This phenomenon demonstrates the critical role of traditional environmental knowledge, land management systems, beliefs, rituals, and taboos in human interactions with environmental disasters.

History, therefore, suggests that residents can adapt to outside disturbances. In doing so, people can refine specific areas of knowledge and methods using local geography. By these means, communities can support the dynamic equilibrium of nature without collapse following some disasters. Researchers have argued that traditional ecological knowledge helps indigenous people continually to adapt to environmental changes and that this information can enhance resilience of individuals and communities through flexible methods of coping with environmental disasters (Boillat & Berkes, 2013; Folke, 2006; Gómez-Baggethun, Corbera, & Reyes-García, 2013; Gómez-Baggethun, Reyes-García, Olsson, & Montes, 2012; Pearce, Ford, Willox, & Smit, 2015; Reyes-Garcia, 2015).

Some studies suggest that culture directly affects recovery processes, including disaster countermeasures (Jang, 2005; Jang & LaMendola, 2006). For example, Rabin (2005) implied that the success of external disaster prevention and relief efforts occurs when the actors understood relevant regional cultures. Cox and Perry (2011) also emphasised the importance of local culture, such as rituals, social organisations and traditional crops, in community recovery, not only as an orienting goal during this recovery, but also as a foundation upon which social capital and community disaster resilience are built and re-built. There is a need from this perspective to consider the dynamics of contextual and cultural factors that influence the disaster recovery process, and this is the aim of the research reported here.

Research methods

In-depth interviews and on-site observations were conducted during 2017–2018 to measure the effect of resilience on post-disaster reconstruction and recovery of the Karamemedesane community, nearly 10 years after the typhoon. Rukai people were asked about their traditional knowledge of disaster prevention, evacuation strategies, land-use practices, patterns of social activity and organisation during disasters. In addition to recovery and reconstruction, villagers were also asked how quickly people became united, adjusted to post-disaster pressures and lifestyle challenges, and returned to daily routines. Participant observations aimed to record the rituals, industries, educational practices, and community reconstruction of the Karamemedesane people.

Snowball sampling was used to identify and recruit key informants to learn about important issues related to post-disaster recovery and reconstruction. Twenty stakeholders were interviewed, including one traditional leader, five political leaders, seven community organisation leaders, two religious leaders, and five villagers. The traditional leader was the chieftain of Karamemedesane who played an important role in facilitating traditional food allocation and production methods such as hunting, gathering, and farming. After the typhoon, the leader's attitude decided the direction of the villagers on the issues of relocation and cultural revival. The five political leaders were the village head of Karamemedesane, the elected representation of Karamemedesane, the mayor of Wutai Township, a councilman of Pingtung County, and the Vice-Director of the Department of Indigenous Peoples of the Pingtung County Government.

These political leaders were in charge of government resource distribution, as well as post-disaster recovery directions, priority, policy and implementation. The community organisations and associations played key roles in promoting social, religious, and economic welfare through grass-roots efforts. Lastly, the five interviewed villagers were members of Karamemedesane's agricultural and marketing group, which united those interested in farm production and marketing. Interviews were held in places such as houses, farmlands, public spaces, offices, and government buildings. Each interview lasted approximately 1–2 h.

The impact of village migration policies and permanent housing measures

After the Morakot disaster, the government implemented several policies for indigenous people, including infrastructure renovation, home redevelopment, and livelihood reconstruction. For infrastructure, there were road and bridge repairs, stream and reservoir dredging, etc. For home redevelopment, relocation of villages and permanent housing occurred. For livelihood reconstruction, several community organisations received funding and social support for sustaining villagers at the relocation area. Among these mandates, villagers thought that the government's relocation policy was too crude and urgent, suppressing opportunities for discussion. This process often resulted in community disputes, affecting unity and coherence.

After Typhoon Morakot, 196 households from Wutai Township moved to LTC, including 73 households from Adiri Village, 39 from Kinulane Village, 59 from Karamemedesane Village, 22 from Kudrengere Village, and 3 from Labuwan Village. Although nearly 200 households were affected, numerous villagers moved to other places. However, a few strong protesters refused to leave their original houses, rather than moving to LTC. Villagers applied for 89 permanent houses in the first phase of relief efforts after Morakot, but only 42 of them won approval by the government and the Tzu Chi Foundation (i.e. the organiser of first-phase construction). Villagers applied for 30 houses in the second phase, but the government and the Red Cross Society of ROC (i.e. the organisers of second-phase reconstruction) approved only 17 of them. In total, only 59 of the 119 applications for permanent housing were approved for Karamemedesane villagers (less than 50%).

This hasty relocation policy caused Karamemedesane people to disperse, rather than consolidating them. To make matters worse, the Rukai and Paiwan were placed in LTC without considering their previous history of enmity. One interviewee, a council member of Pingtung County, suggested that relocation was a genocide policy, completely undermining the integrity and uniqueness of Rukai.

After allocating permanent houses to qualified applicants, the government demanded that people leave their original homes within three months and prohibited them from rebuilding at that location. According to the Distribution Principles of Permanent Houses issued by the Pingtung County Government,

> Party C (i.e. the disaster-affected people), his or her spouse, and cohabitant lineal descendant (s) shall agree to move out of the original place of residence within 3 months after the closing date or prior to the final move-out date specified by Party B (the government) and shall not return to live at or develop the property of the original place of residence. The Pingtung County Government will conduct inspections at irregular intervals regarding household registration status to verify that the actual occupant number is consistent with the number specified in the original application, and inspection results shall be considered as the basis for review and reallocation of property.

Affected villagers would lose their rights to live in permanent housing if they did not change their household registrations within three months. According to the Vice-Director of the Department of Indigenous Peoples of the Pingtung County Government:

> Once the affected villagers move away from the original places, technically we don't want them to move back. To be more specific, the affected people are not allowed to move back there (to their original places of residence). We already asked villagers to move out, so how can we let them keep their original household registration?

One affected villager remarked, 'If we change our household registration, doesn't it mean that we give up our rights to move back home?'

In contrast to migration or temporary departure strategies previously adopted by villagers who had experienced land shortage caused by environmental disasters, this government policy demanded that people live at specific places and prohibited them from leaving without permission. It also inhibited villagers in subsequently changing their community or residence in response to further environmental and social situations. Moreover, residential areas allocated to villagers were small because the LTC was limited in size. No extra spaces were included around the main buildings, restricting their occupants' activities. A few households were assigned to 46 square metre houses with only two rooms, forcing some families to sleep on the floor in their living rooms. Interviewee A4 said:

> Before we moved in, the government kept urging us to move to LTC and boasting that the permanent houses were much more spacious than our current places. It turned out that these permanent houses are cramped and the surrounding areas are small, which have limited our lifestyles. The government hasn't given us enough living space or offered us jobs, so we don't know what to do in the future. This is not what the government promised.

Villagers could tolerate small living areas because the houses only served as places to sleep. However, the larger challenges were livelihood reconstruction and job opportunities. After the residents moved to LTC, the government did not provide them with any farmland. Although some villagers found work in the cities, numerous middle-aged and older people did not possess the required knowledge or skills for modern jobs. Consequently, those individuals experienced considerable pressure on their day-to-day lives and finances. Disasters not only cause damage to the environment, buildings and physical facilities, but also aggravate personal problems, especially for those with pre-existing social and financial limitations. Disasters often result in undesirable social and cultural conditions for villagers, thus increasing vulnerability. Simultaneous interaction of these factors can produce considerable social, economic, cultural, and day-to-day-life impacts. Hence, identifying solutions for post disaster recovery, followed by the mechanisms and motives that facilitate this recovery is a crucial aspect of disaster research. As Oliver-Smith (2001) wrote:

> ... many people imagine that, first, a disaster occurs; next, the world mobilizes to ... assist in the distribution of tents, blankets, medicine, and food. I don't do that ... The reconstruction phase is the longest, most expensive, most complex and politically volatile of the disaster, which is highly related to the ethnic group, class, gender, and differential patterns of aid distribution and deeply influences the affected people. (Oliver-Smith,2001, pp. 111–112).

Traditional agriculture and food crops' security

Compared to other indigenous groups who had given up traditional farming altogether, the Rukai people kept their practice of slash-and-burn farming until the 1980s when they started to develop homestay businesses for tourists. Notwithstanding the change over many years, a majority of the Rukai people remained in farming and continued to rely on traditional agriculture. As part of this practice, in Taiwan, indigenous people traditionally grow red quinoa (*quinoa Chenopodium formosanum*) mainly for ritual purposes, while simultaneously cultivating other food crops including foxtail millet, taro, and sweet potatoes. In addition to culture, traditional food crops play an important role for Rukai people who encounter environmental disasters. Baked taros (*kurai*) are dried for convenient long-term storage and transportation. *Kurai* is a valuable food source for Rukai people in times of famine. Interviewee R said that crops such as millet, peanuts, taros, and pigeon peas were essential when typhoons struck their mountainside village. Hence, it was a requirement to plant these crops in the same field to minimise risk.

After harvesting, crops were processed and stored in a *pulutu* (a cylinder enclosure fabricated from a hollowed camphor tree) or *tausulu* (a square room constructed from piled slates). Dried food can be stored for 3–5 years, but villagers cleaned or replaced the crops on a regular basis to avoid moulding and to ensure safe consumption, as needed. When a typhoon struck, villagers distributed their dried taro and other food products among themselves. This countermeasure became a life experience, promoting culture through mutual aid (Taiban, 2012).

If a typhoon destroyed roads or bridges that led to other towns, then villagers lost access to food products and other supplies. Typhoons often cause long-term power outages. Without electricity, water, and gas, villagers would exhaust their food supplies, thereby facing a survival crisis. Under such circumstances, villagers would have to rely on airdrops or evacuation. Although traditional survival measures are no longer in use, villagers often experience panic and helplessness when typhoons strike their village.

Villagers relied on traditional food sources, such as red quinoa, foxtail millet, taros, sweet potatoes, and pigeon peas since these crops could be grown with small amounts of water and fertiliser, tolerate high temperatures, and adapt to climate change (Newton et al., 2011; Sambo, 2014). Traditional farming practices and geographical knowledge of villagers was crucial for community resilience since it determined sustainability of local food systems and community livelihoods (Ba, Lu, Kuo, & Lai, 2018; Shava, O'Donoghue, Krasny, & Zazu, 2009).

In 2010, when the Karamemedesane community moved to LTC, the government provided accommodation, but no land for farming. Older people became bored, and many passed whole days sitting in front of their houses or watching television. Interviewee A5 argued, 'We'd [the villagers] rather live in ordinary houses with farmland. Land helps us survive and give(s) us hope. Without land, we can't produce food for ourselves. We become uncertain and the village elders fall ill'. In fact, numerous older people lost the will to live during the transition process following community migration and eventually died after experiencing deep depression.

Growing cash crops

Under pressure from the villagers, the Pingtung County Government finally rented a 20-hectare plot of land from the Taiwan Sugar Corporation in 2010 for LTC residents to grow crops.[1] However, villagers could not make their own decisions; the county government made them grow specific crops by collective farming practices. The government claimed that villagers should cultivate 'suitable' crops and in doing so, become familiar with local agricultural markets. The first plan required villagers to cultivate 2 hectares of sweet corn. It failed miserably due to implementation by various governmental departments and poor soil quality. This inefficiency produced a shortage of labourers for planting, resulting in low participation for this initiative.

After the failure of sweet corn, the county government consulted experts and launched a plan in 2012 to grow pitaya (commonly known as dragon fruit). The intention was for pitayas to become the major crop in LTC. The government claimed that pitayas were ideal for villagers to grow because the crop was easy to manage, thereby saving time and lowering costs. Additionally, pitayas were durable and had a high sugar content. A registration process determined the allocation of field shares. Nearly 40 households registered initially, but the amount of farmland was not sufficient to meet demand. Finally, shares of only 0.3 ha each were allotted to 33 households.

The government then invited pitaya vendors to assist villagers for 1 year of cultivation, promising 10 years of acquisition at a guaranteed price. The export market for pitayas included Canada, Japan, and South Korea. Despite these considerations, interviewee A6 expressed some dissatisfaction: 'Why sweet corns and pitayas? What we actually want to grow are our traditional food crops … '. Traditional food crops consisted of foxtail millet, sweet potatoes, taro, and red quinoa, but these crops were not approved by those in charge. Government officials argued that traditional food crops were not staple items and would not be marketable because of low production.

The vice chairperson of LTC management committee asserted that 'The proposed initiative requires villages grow preselected crops, and the villagers have to work regular schedules like corporate employees'. This policy was opposite to what the LTC villagers wanted, which was for them to grow the crops they wanted such as millet, taro or red quinoa, just as they did in their original farmland. One of the villagers commented: 'Why do we have to grow pitayas? Pitayas are not one of our traditional food crops. We don't need pitayas during the Harvest Festival. Why can't we grow millet?' Many LTC villagers believed that the government wanted villagers to grow cash crops to solve economic problems. In doing so, it neglected the continuation and preservation of culture.

To solve the problem of insufficient land, the government provided an extra 6 hectares of farmland for villagers. However, 18 of the 40 households who rented plots in this area wanted to terminate their leases because their farmland was 5 km away from LTC, requiring villagers to travel by cars or scooters. Because most farmers were over 50 years old, the journey between LTC and the farmland was inconvenient and unsafe. Moreover, the land had been vacant for 5–6 years before it was acquired, making it costly and time-consuming for villagers to rehabilitate.

Returning to tradition

After the failed cultivation of sweet corn and pitayas, villagers asked government officials to grow red quinoa as the major LTC crop beginning in August 2013. Villagers regard red quinoa and foxtail millet as sacred crops, hence valuable. For example, the use of red quinoa is limited to ritual offerings and as food for the elderly, distinguished guests, sick people, and pregnant women. Long ago, villagers consumed red quinoa as a food source, but sold it infrequently to non-villagers due to its scarcity. According to Tsai (2009), the nutritional value of red quinoa is higher than that of regular quinoa. For example, it is equivalent to beef in protein content, delivers 50 times as much calcium as rice, and has seven times the dietary fibre of sweet potatoes. If consumed with the chaff, red quinoa provides antioxidants. This crop is an important ingredient in noodles and cakes, and can be used as nutritional supplements, cosmetics, and skin care products due to its high anthocyanin content. Therefore, red quinoa has a large economic potential.

In 2013, the county government facilitated a contract between LTC communities and KULLKU, a corporation that specialised in developing red quinoa farms. Its purpose was to help villagers learn commercial practices related to red quinoa, including cultivation, threshing, screening, pest control, sampling and inspection, packaging, and marketing.

After villagers left the mountainous areas, their cost of living increased owing to the costs of water, food, and transportation, the last because they depended on purchases in markets in towns.

Agriculture, therefore, became more than a means of providing food; it became a primary source of household income. To earn sufficient profits, villagers had to consider market characteristics, identify buyers, and sell products at desirable prices. After launching the contract-based farming scheme, villagers began to transport red quinoa from the farms to LTC. Monetary transactions occurred after the crops were dried and weighed. In some cases, villagers would receive same-day, cash payments. This method of operation enhanced trust with the KULLKU company, resulting in a willingness to grow more crops. In 2013 (first stage), approximately 30 villagers participated in contract-based farming and produced an output of 1,500 kg. KULLKU acquired the products at $4.3 USD per kg. Initially, the demand for red quinoa was high. Various brokers visited LTC and offered to buy the crops at high prices. Villagers became motivated and increased its output to 3000 kg during the second stage.

The county government subsidised rental of the Spirit Farmland for the first 2 years of use. However, in August of the third year, LTC residents started to pay rent ($2000 USD per hectare). This decision caused villagers to become strategic in their crop selection and selling practices. One of the villagers reasoned: 'Still, we must have regular incomes, meaning that struggling alone is not the solution. If we establish a production and marketing group, we can achieve better operations and integrate crop production with marketing. Thus, incomes will go up'.

In January 2013, LTC residents established the First Production and Marketing (PM) Group of Special Crops (red quinoa, chili peppers, and other vegetables) at Wutai

Township, Pingtung County. The work of growing and harvesting red quinoa enriched the life of LTC villagers who joined the PM group. At last, villagers were able to recover some of their original lifestyle.

December 2013 marked the first harvest of red quinoa. Shares of the 6 hectares used to produce red quinoa were assigned to PM group members. Traditional farming methods were used for harvesting. For example, group members cooperated with other households to harvest in shifts, based on a schedule for each crop. After harvest, households dried and threshed their crops. Mr. K from Karamemedesane estimated that the harvest of unthreshed red quinoa was about 10 tons, reflecting a bountiful harvest for that year. Mr. K added that the villagers employed eco-friendly farming techniques and cultivated certified products that were free from pesticide residues. Each commercial unit of red quinoa was vacuum-packed and weighed 400 catties (240 kg). Delivery was free for the orders of $\geq$10 packs.

Since red quinoa yielded high prices in the markets, more farmers, including non-indigenous people, started to grow it. Sales dropped as supply exceeded demand. When the sale of red quinoas decreased, the PM group, led by Mr. K, instructed unemployed villagers and homemakers to categorise, classify, thresh, and screen red quinoa, and perform vacuum packing. Additionally, villagers learned how to use social media platforms such as Facebook and LINE to sell products. Mr. K explained that selling red quinoa was similar to other farm products. However, red quinoa did not spoil, if it was dried and stored properly. Farmers did not sell red quinoa when the price dropped because this practice not only reduced market value, but also prevented a price rebound. When the market price was low, farmers established PM groups to participate in contract farming, thereby ensuring sales.

But the story here takes another turn. Mr. K was one of the few Karamemedesane people who opposed relocation after the typhoon. LTC villagers faced the challenge of insufficient farmland, coupled with high rents. As a result, about 10 of the villagers started growing crops on the mountainside farmlands, going back and forth to do so, despite the possible risk from future typhoons to their crops there. Most of these individuals were 40–60 years old, but some were older than 70. Two-thirds of them were females, a departure from Rukai tradition. Traditionally, Rukai men were responsible for farming and hunting, while women stayed home to take care of children and livestock. After relocation to lowlands, most male villagers left their permanent houses to earn money. Thus, the farming work which requires hard labour fell on the women. In 2015, Mr. K founded the Second PM Group consisting of these 10 LTC residents and 10 mountainside villagers (those living in the safe zone) and motivated them to plant red quinoa, foxtail millet, taro and tree beans for the marketplace, not simply for personal consumption. Elders taught young people about traditional farming practices. Young people combined this knowledge with some new techniques, such as mechanical farming tools, food processing and vacuum packing.

Through their own efforts, members of the second PM group created innovative products made from red quinoa such as cakes, noodles, crackers, and tea. In September 2017, the group decorated the abandoned Karamemedesane Elementary School building and founded the Academy of Special Rukai Crops (ASRC). Members opened a village bakery to preserve their traditional ecological knowledge by growing crops suitable for the bakers to use. Every member received an equal share of profits, a practice that reflects Rukai culture. Overall, modern techniques of food processing, product development, and marketing improved the village economy, while addressing food safety issues. Growing traditional crops under Mr. K's leadership represents a win-win situation for both LTC residents and mountainside villagers.

Small changes for transformability, rooted in land

The farming culture of villagers was revitalised after seeing the importance of traditional crops. For example, efforts were made to preserve essential Rukai agricultural knowledge by establishing two PM groups and the ASRC. Land enabled Karamemedesane villagers to find several opportunities for revival and transformation following the onset of the environmental disaster.

Folk et al. (2010) stressed that one important aspect of resilience was transformability, the capacity to cross thresholds into a new development trajectory. Transformation draws on resilience from multiple scales, thereby making use of a crisis as a window of opportunity for novelty and innovation, and recombining sources of experiential knowledge to navigate social-ecological transitions. Villagers recovered by using the power of the land. Small-scale changes enabled resilience at larger scales. In this study, these changes included applications of traditional ecological knowledge (e.g. red quinoa), leadership (e.g. Mr. K), social networks and social learning (e.g. PM groups) as well as innovation (e.g. the village bakery). Each of the factors allowed dynamic shifts for transformability (Berkes, Colding, & Folke, 2003; Folke, 2006; Folke et al., 2010). This study showed that resilience is rooted, maintained and revived through land-based practices.

According to Huang (2018a), a broader understanding of disaster justice is needed. Existing literature tends to devalue bottom-up, community-based reactions to disasters, yet many state-recognised definitions exclude the socioeconomic and environmental contradictions that caused people to become vulnerable. The focus should be on socio-spatial production of risk, rather than simply using disasters as examples for planning and recovery in emergency response situations. This definition should include state-led relocation projects as examples of human-made disasters, especially those which deprive local people the right to practice self-governance and land management as they have done in the past. In this study, the Karamumdisane community was sub-divided because of Typhoon Morakot. Half of its residents were forced to relocate to LTC because their houses were declared unsafe, but the other half remained on the mountainside. Although growing traditional crops increased LTC villagers' resilience, the challenge of insufficient farmland, coupled with high rental prices at lower elevation increased their vulnerability. LTC residents solved these problems by cooperating with villagers who lived in safe areas on the mountainside. Cooperation required social capital, social networks, social learning, leadership and innovation. These capacities enabled villagers to re-orient themselves in navigating displacement, self-governing post-disaster recovery and regaining knowledge in land management. Each was derived from small changes in land-based practices that resulted in increased resilience.

The success of Karamemedesane in restoring traditional agriculture has given inspiration to other Rukai villages faced with relocation such as Adiri, Kinulane and Labuwane. The trend of 'going back to homeland to grow traditional crops' also drew attention from a nearby university to develop 'under-forest economy' (also called Agroforestry). This practice seeks to cultivate traditional crops, such as red quinoa, millet, melons, taro, as well as gold-line lotus and mushrooms under the forest cover to produce economic benefits. Recently the mayor of Wutai Township established 'Wutai Township Natural and Cultural Landscape Area,' which has a legal basis for promoting eco-tourism, under-forest economy, as well as other sustainable practices.

Conclusions

The plan to relocate Rukai villagers far away from their ancestral homelands after Typhoon Morakot resulted in a failed government policy. It did not account for safety, accommodation, farming practices, traditional crops, jobs, emotional attachments to place, etc. Rukai have lived in mountainous areas all of their lives, developing some unique forest-related connections, myths, and traditional taboos. After moving to a lower elevation, much of their traditional knowledge became obsolete. Many of the villagers struggled to adapt their lifestyle and customs to a new environment, causing severe emotional distress, and nearly spiraling into social disintegration. As villagers left the mountainsides, their land became useless.

Post-disaster policies must be rooted in the daily habits of the residents, such as traditions in architecture, farming, industry, and diet. In this study, Karamemedesane villagers did not derive the energy of spiritual warmth and cultural recovery from safe and durable houses provided by the government. Instead, they healed their wounded hearts by growing traditional crops and adapting through manual labour. This study suggested that community resilience is rooted in the land and culture of these indigenous people. Government policies should consider ways to promote small and culturally relevant changes to foster resilience and transformability using new development trajectories.

Future research should focus on understanding indigenous communities, especially in transforming social and political relationships between villagers and government authorities. What kind of land-based, culturally relevant, post-disaster policies can foster new development trajectories for indigenous communities? It is our hope to include, represent, and connect the voices of indigenous communities in post-disaster recovery, so as to aid resilience and promote transformability.

Note

1. The 20 hectares of land were adjacent to the Linluo Interchange. Of this total, 11 hectares were designated as a Special Crops Area, 6 hectares were designated as Spirit Land, and 3 hectares were set aside for other purposes. Two hectares of land in the Special Crops Area located near Ailiao Military Camp were used for cultivation of sweet corn.

Acknowledgements

We also would like to thank Dr Mark Morgan for his comments and editorial work to improve the quality of this manuscript.

Disclosure statement

No potential conflict of interest was reported by the authors.

Funding

We would like to thank the Ministry of Science and Technology of the Republic of China (Taiwan) for financially supporting this work under Contract Nos. MOST-106-2420-H-214-001-MY3.

References

Ba, Q.-X., Lu, D.-J., Kuo, W. H.-J., & Lai, P.-H. (2018). Traditional farming and sustainable development of an indigenous community in the mountain area – a case study of Wutai village in Taiwan. *Sustainability, 10*(10), 3370.

Berkes, F., Colding, J., & Folke, C. (Eds.). (2003). *Navigating social-ecological systems: Building resilience for complexity and change*. Cambridge, UK: Cambridge University Press.

Blaikie, P. M., Cannon, T., Davis, I., & Wisner, B. (1994). *At risk: Natural hazards, people's vulnerability, and disasters*. London: Routledge.

Boillat, S., & Berkes, F. (2013). Perception and interpretation of climate change among Quechua farmers of Bolivia: Indigenous knowledge as a resource for adaptive capacity. *Ecology and Society, 18*(4), 21. doi:10.5751/ES-05894-180421.

Brown, D., & Kulig, J. (1996). The concept of resiliency: Theoretical lessons from community research. *Health and Canadian Society, 4*, 29–52.

Cohen, A. (1974). *Two dimensional man: An essay on power and symbolism in complex society*. London: Routledge and Kegan Paul.

Cox, R. S., & Perry, K. M. (2011). Like a fish out of water: Reconsidering disaster recovery and the role of place and social capital in community disaster resilience. *American Journal of Community Psychology, 48*(3-4), 395–411.

Folke, C. (2006). Resilience: The emergence of a perspective for social-ecological systems analyses. *Global Environmental Change, 16*(3), 253–267.

Folke, C., Carpenter, S. R., Walker, B., Scheffer, M., Chapin, T., & Rockström, J. (2010). Resilience thinking: Integrating resilience, adaptability and transformability. *Ecology and Society, 15*(4), 20. Retrieved from http:// www.ecologyandsociety.org/vol15/iss4/art20/

Gómez-Baggethun, E., Corbera, E., & Reyes-García, V. (2013). Traditional ecological knowledge and global environmental change: Research findings and policy implications. *Ecology and Society, 18* (4), 72. doi:10.5751/ES-06288-180472.

Gómez-Baggethun, E., Reyes-García, V., Olsson, P., & Montes, C. (2012). Traditional ecological knowledge and community resilience to environmental extremes: A case study in Doñana, SW Spain. *Global Environmental Change, 22*(3), 640–650.

Hewitt, K. (Ed.). (1983). *Interpretation of calamity: From the viewpoint of human ecology*. Winchester, MA: Allen & Unwin.

Holling, C. (1973). Resilience and stability of ecological systems. *Annual Review of Ecology and Systematics, 4*, 1–23.

Huang, S. M. (2018a). Understanding disaster (in)justice: Spatializing the production of vulnerabilities of indigenous people in Taiwan. *Environment and Planning E: Nature and Space, 1*(3), 382–403. doi:10.1177/2514848618773748.

Huang, S. M. (2018b). Heritage and postdisaster recovery: Indigenous community resilience. *Natural Hazards Review, 19*(4), 05018008. doi:10.1061/(ASCE)NH.1527-6996.0000308.

Jang, L. J. (2005). *The 921 earthquake: A study of the effects of Taiwanese cultural factors on resilience* (Unpublished doctoral dissertation). Denver: University of Denver.

Jang, L., & LaMendola, W. (2006). The Hakka spirit as a predictor of resilience. In D. Paton & D. Johnston (Eds.), *Disaster resilience: An integrated approach* (pp. 174–189). Springfield, IL: Charles C. Thomas Publisher.

Nelson, D. R., Adger, W. N., & Brown, K. (2007). Adaptation to environmental change: Contributions of a resilience framework. *Annual Review of Environment and Resources, 32*, 395–419.

Nelson, D. R., West, C. T., & Finan, T. J. (2009). Introduction to 'In focus: Global change and adaptation in local places'. *American Anthropologist, 111*(3), 271–274.

Newton, A. C., Flavell, A. J., George, T. S., Leat, P., Mullholland, B., Ramsay, L., ... Bingham, I. J. (2011). Crops that feed the world 4. Barley: A resilient crop? Strengths and weaknesses in the context of food security. *Food Security, 3*, 141–178.

Oliver-Smith, A. (2001). Anthropology in disaster research and management. *National Association for the Practice of Anthropology Bulletin, 20*(1), 111–112.

Oliver-Smith, A., & Hoffman, S. (1999). Anthropology and the angry earth: An overview. In A. Oliver-Smith & S. Hoffman (Eds.), *The angry earth: Disaster in anthropological perspective* (pp. 1–16). New York, NY: Routledge.

Paton, D., & Johnston, D. (2001). Disasters and communities: Vulnerability, resilience and preparedness. *Disaster Prevention and Management: An International Journal, 10*(4), 270–277.

Pearce, T., Ford, J., Willox, A. C., & Smit, B. (2015). Inuit traditional ecological knowledge (TEK) subsistence hunting and adaptation to climate change in the Canadian Arctic. *ARCTIC, 68*(2), 233–245.

Pelling, M. (2003a). *The vulnerability of cities: Natural disasters and social resilience*. London: Earthscan.

Pelling, M. (2003b). The political ecology of flood hazard in urban Guyana. In K. S. Zimmerer & T. J. Basset (Eds.), *Political ecology: An integrative approach to geography and environment-development studies* (pp. 73–93). New York, NY: Guildford Publications.

Rabin, C. L. (Ed.). (2005). *Understanding gender and culture in the helping process: Practitioners' narratives from global perspectives*. Belmont, CA: Thomson/Wadsworth.

Renaud, F., Bogardi, J., Dun, O., & Warner, K. (2007). *Control, adapt or flee: How to face environmental migration. InterSecTions, 5*. Bonn, Germany: UNU-EHS.

Renaud, F. G., Dun, O., Warner, K., & Bogardi, J. (2011). A decision framework for environmentally induced migration. *International Migration, 49*(1), e5–e29.

Reyes-Garcia, V. (2015). The values of traditional ecological knowledge. In J. Martinez-Alier & R. Muradian (Eds.), *Handbook of ecological economics* (pp. 283–306). Cheltenham, UK: Edward Elgar.

Sambo, B. E. (2014). Endangered, neglected, indigenous resilient crops: A potential against climate change impact for sustainable crop productivity and food security. *IOSR Journal of Agriculture and Veterinary Science, 7*(2), 34–41.

Shava, S., O'Donoghue, R., Krasny, M., & Zazu, C. (2009). Traditional food crops as a source of community resilience in Zimbabwe. *International Journal of African Renaissance Studies, 4*(1), 31–48.

Taiban, S. (2008). The division and reconstruction of traditional territory: Re-examining human-land configuration and spatial change of Kucapungane. *Journal of Archaeology and Anthropology, 69*, 9–44.

Taiban, S. (2012). Disaster, relocation and vulnerability: The case study of Kucapungane. *Taiwan Journal of Anthropology, 10*(1), 51–92.

Taiban, S. (2013). From Rekai to Labelable: Disaster and relocation on the example of Kucapungane, Taiwan. *Anthropological Notebooks, 19*(1), 59–76.

Tsai, P. J. (2009). Nutrition and application of red quinoas. *Agricultural World, 307*, 18–23.

Post-disaster communalism: land use, ownership, and the shifting 'publicness' of urban space in recovery

Elyse M. Zavar and Ronald L. Schumann III

ABSTRACT

Disasters disrupt the functions and meanings of urban spaces, often prompting changes in their ownership, management, and use. This study examines two such disaster events: riverine flooding in Lexington, Kentucky, and Hurricane Katrina in coastal Mississippi. After each event, we observe the emergence of quasi-public spaces during long-term recovery, where residents generally perceive an increase in the 'publicness' of formerly exclusive spaces. This trend more closely mirrors public-private hybrid spaces discussed in urban geographic literature rather than strict privatisation as observed in most post-disaster studies. Terming this shift from private to greater public use *post-disaster communalism*, we employ data from open-ended surveys, photovoice, and resident interviews to describe the phenomenon. Using discourse analysis and grounded theory, we explore residents' perceptions of access and belonging in these quasi-public urban spaces. Findings show how different policy instruments (public buyouts versus insurance rate changes) contribute to similar land uses and perceptions of access. Building upon previous human geography research, we discuss five broad mechanisms that collectively generate communalism in our study sites. Finally, we consider the competing implications of post-disaster communalism for promoting both ideals of public space and of community resilience to future disasters.

1. Introduction

Within the current neoliberal political environment, which has encouraged privatisation of services and activities once the responsibility of municipal government (Harvey, 2005), a new trend is emerging in post-disaster communities: private land transitioning to public use. As disasters transform the cultural and physical landscapes of communities, first through destruction by hazard agents and second through the politics of post-disaster reconstruction and risk mitigation, the utility and ownership of land within the disaster zone can also change (Geipel, 1991; Haas, Kates, & Bowden, 1977; Hagelman, Connolly, Zavar, & Dahal, 2012; Rosen, 1986; Zavar, Hagelman, Lavy, & Prince, 2017). Mitigation activities, such as levee construction or the removal of structures in high-risk areas, directly transform the physical appearance of spaces, thereby shifting how land is used and by whom. Modifications to zoning laws or insurance pricing structures instituted during community recovery also influence how property owners elect to rebuild, bringing about different post-disaster land use patterns. These dynamics produce quasi-public spaces after disasters—spaces that may be publicly or privately owned, but which function and are perceived as public.

In this study, we investigate this emergence of quasi-public spaces, which we term *post-disaster communalism*, after two events. The first case explores floodplain buyouts in Lexington, Kentucky. This mitigation strategy transferred high-risk private land to public ownership (Zavar, 2015). The second case examines the recovery experience of coastal Mississippi after Hurricane Katrina where the post-disaster landscape invited greater public use of previously exclusive spaces. Triangulating between these cases, we introduce the concept of communalism, analyse residents' perceptions of communalism, discuss mechanisms that influence public access, and consider communalism's impacts on the ideals of public space and community disaster resilience.

2. Policies shaping public-private hybrid spaces

Neoliberal policies frequently reduce the publicness of public space (Mitchell, 2010), blurring the distinction between public and private. Public-private hybrid spaces are identified 'by the partial or complete transfer of state or local rights to private or commercial actors as well as by the reduction or even loss of public control' (Nissen, 2008, p. 1131). Hybrid spaces obscure rights to access and feelings of belonging. Exclusion and belonging are demonstrated through displays of power ranging from overt to tacit (Trudeau, 2006). For instance, zoning ordinances represent an overt, explicit, and government-sanctioned example of who and what is permitted on the landscape (McCann, 1997). By contrast, landscape aesthetics subversively convey norms of belonging and exclusion (Duncan & Duncan, 2001). Regardless of how the politics of belonging, and consequently exclusion, manifest, the result is that established groups (i.e. the affluent, dominant) maintain access while others' access is limited or denied (Sibley, 1995).

Reasons for public-private hybridisation are often financial. Many cities cannot afford the costs associated with maintaining public space, so private partnerships reduce financial burdens (Zukin, 1995). Created to reduce the municipal budget, New York City relies on privately-owned public spaces (POPS). POPS are legally required to be open to the public during certain hours yet are owned and maintained by private entities (Németh, 2009). POPS range from high-functioning public spaces to those managed in a way that excludes the public (Kayden, 2000).

2.1. Disaster-induced changes to the urban environment

Limited research examines how disasters impact issues surrounding public-private spaces. Extant scholarship explores how existing neoliberal politics compound or exploit disaster relief and recovery (e.g. Klein, 2007; Trujillo-Pagan, 2012) rather than how disasters generate changes in public-private management of urban space. Scholars have documented that disasters amplify trends that existed pre-disaster, which then alter post-disaster physical and cultural landscapes (Hagelman et al., 2012; Pais & Elliott, 2008). For example, a community experiencing economic growth prior to a disaster event will experience accelerated growth afterward during the time-compressed recovery period; these conditions encourage uneven redevelopment and induce reorganisation of the community's sociodemographic layout (Pais & Elliott, 2008).

Post-disaster economic growth or decline affects what is rebuilt, thus altering land use and landscape composition. Increased urbanisation is a commonly cited post-disaster land use change (Geipel, 1991; Haas et al., 1977; Zavar et al., 2017). Often the physical footprint of a community expands as buildings develop on the periphery and later are included within the recovered community boundaries (Hagelman et al., 2012). Conversely, disaster-impacted areas facing issues of housing unaffordability due to market conditions or insurance rate hikes can experience checkerboarding, whereby vacant parcels remain interspersed among rebuilt structures for years post-disaster (Cutter et al., 2014). These changes during the disaster reconstruction period influence public/private ownership and management of land.

Accompanying post-disaster changes in physical spaces are changes in community members' perceptions of impacted/altered landscapes. Disasters disrupt underlying bonds, or place attachments, people have with the places they inhabit (Brown & Perkins, 1992). Catastrophic disasters render neighbourhoods unrecognisable, generating feelings of homesickness among survivors (Erikson, 1976). The fear of losing one's place to a disaster spurs acts that reinforce individual or group claims to physical spaces in recovering communities. These acts of territoriality seek to: (a) demarcate/defend spaces, (b) maintain the social order in stressful times, and (c) provide an external outlet for identity displays (Brown, 1987). Notions of ownership, exclusiveness, and control over the appearance and uses of space undergird territoriality (Brower, 1980). Therefore, we investigate both the creation of quasi-public spaces in post-disaster urban environments and residents' perceptions about their 'publicness.'

2.2. Policies altering post-disaster housing & land use

During the 1990s, changes to United States federal disaster policy placed greater emphasis on hazard mitigation activities, thereby altering floodplain management strategies pre- and post-disaster. Property acquisitions, or buyouts, came into widespread use to eliminate structures from floodplains and remove high-risk properties from private real estate markets altogether. Catastrophic flooding across the Midwestern U.S. in 1993 prompted amendments to the Stafford Act, which increased federal funding allocations for post-disaster buyout programmes (Godschalk, Beatley, Berke, Brower, & Kaiser, 1999). Additionally, the National Flood Insurance Program (NFIP) Reform Act of 1994 created hazard mitigation funding aimed at relocating residents out of flood prone areas prior to disaster events; the act further mandated that all private lenders require flood insurance for the life of each federally backed home mortgage loan (Godschalk et al., 1999).

Enacted at the federal level, these policies aimed to reduce federal disaster recovery expenditures, indemnify property owners for losses, and guide local planning activities. While these policies succeeded in shaping municipal planning strategies nationwide, they also interacted with local housing markets, at times producing unintended consequences for housing affordability and land use patterns. Previous scholarship identifies a myriad of factors that influence housing value. Parcels in close proximity to open space can increase home values in the U.S. (e.g. Bolitzer & Netusil, 2000); however, the impact of the NFIP on home values vary widely from place to place (Montz, 1987; Bin, Crawford, Kruse, & Landry, 2008). Therefore, our case studies detail the localised effects of these interactions between policies and markets, including their impacts on the visual landscape itself.

3. Methods

To understand how disasters impact public space, we utilised two datasets collected independently. The first dataset, a mixed-methods case study, consisted of field observations, a mail survey, and semi-structured interviews with residents living in neighbourhoods that experienced floodplain buyouts. Field observations included repeat site visits during 2012–2013 and a six-week field study over the summer of 2013. Surveys were mailed to 130 households in the two neighbourhoods of Lexington, Kentucky, following Dillman's (1978) Total Design Method in 2013. This study focuses on the open-ended responses including unsolicited comments written on the survey by participants (combined survey response rate of 19.09 percent; Zavar, 2015). During summer 2013, sixteen follow-up interviews were conducted with residents who self-identified on the surveys (Rubin & Rubin, 1995). An inductive coding structure was developed in the software programme NVivo (2012) based on the open-ended survey question responses, interview transcripts, and field notes. To analyse residents' perspectives on the management of the buyout and to interrogate issues related to the utility of the open space (Zavar, 2015; Zavar, 2016), we employed a Critical Discourse Analysis (Fairclough, 1995).

The second dataset derives from a study on long-term recovery after Hurricane Katrina in coastal Mississippi. Following photovoice protocol (Schumann, Binder, & Greer, 2019), during fall 2013, thirty-four resident-survivors of Katrina were asked to photograph objects, people, landmarks, and scenes within their communities that illustrated recovery in progress or that previously occurred (Schumann et al., 2019). Afterward, semi-structured interviews exploring the photos were conducted with participants. Inductive Grounded Theory method was applied to discern meanings of recovery from the interview data (Charmaz, 2006). These data, along with accompanying participant-authored photos, researcher-authored photos, and field notes from nine weeks of fieldwork in 2013 and 2014, document perceptual changes in ownership and access to spaces in two Mississippi towns. Together, these datasets provide a robust account of post-disaster recovery. They offer two independent case studies of different locations impacted by separate disasters at different time periods, thus increasing this study's reliability and validity.

3.1. Study sites

Episodic flooding plagued the Wolf Run watershed in Lexington, Kentucky since development in the mid-twentieth century. In spring of 1997, the watershed experienced record-breaking floods that prompted federally-funded buyouts within the watershed (Zavar, 2015). In this study, we focus on the experience of the Port Royal neighbourhood and the adjacent private street, The Lane. The neighbourhood of Port Royal contains 99 properties following the buyout of 12 properties; The Lane contains 31 properties following the buyout of one property (Figure 1). The buyout properties in both neighbourhoods primarily consist of passive open space characterised by mowed, vacant lots except for a 12-by-

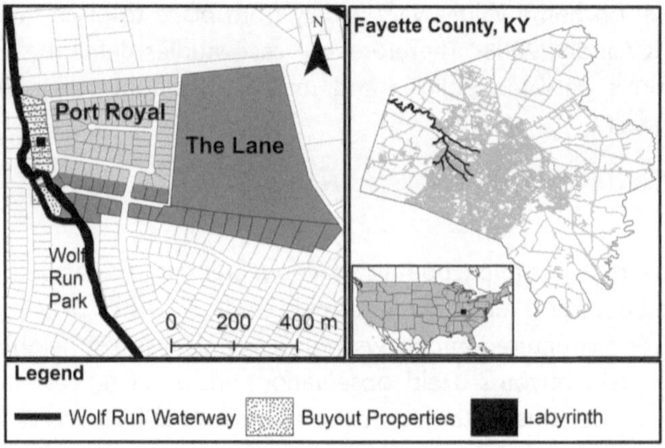

Figure 1. Port royal neighbourhood and the lane map.

12-meter cement slab located in Port Royal. The cement slab is a labyrinth with an imprinted maze pattern that leads to a bench at the centre; the site is intended for contemplation (Zavar, 2016).

Through the voluntary buyout programme, homeowners were compensated the pre-flood value of their properties, and in turn, the Lexington Fayette Urban County Government (LFUCG) gained ownership of the properties and demolished the existing structures to convert the land to open space. In Lexington as elsewhere, buyout properties purchased through federal cost-share programmes like the Hazard Mitigation Grant Program must remain open space in perpetuity, which permanently removes land from private ownership and occupation (FEMA, 1998).

The second study site consists of locations impacted by Hurricane Katrina in two Mississippi beachfront towns: Bay St. Louis and Pass Christian (Figure 2). These municipalities along the western section of Mississippi's coastline received the highest storm surge heights (7-10 m) during the hurricane (Knabb, Rhome, & Brown, 2005). This urbanised coastline had few undeveloped parcels in 2005 when Hurricane Katrina made landfall (Colten & Giancarlo, 2011). Recent hurricanes—the 1947 Hurricane, Hurricane Camille (1969), and Hurricane Katrina (2005)—led to significant changes in development, but only marginal improvements to mitigation practices (Colten & Giancarlo, 2011). Since Katrina, the Great Recession, the Biggert-Waters Flood Reform Act (2012), prohibitive flood and wind insurance rates, and limited reconstruction of utilities have stymied rebuilding (Cutter et al., 2014). Today, the immediate coastline remains a checkerboard of mowed vacant lots, rebuilt raised homes, and overgrown abandoned parcels.

4. Findings

From these cases, we observed instances of expanded public access under a variety of ownership conditions; we term this *post-disaster communalism*. Drawing on the word communal, which evokes notions of collective use and shared property, post-disaster communalism describes a phenomenon whereby the perceived 'publicness' of formerly exclusive spaces increases after a disaster event. Figure 3 contains a framework of

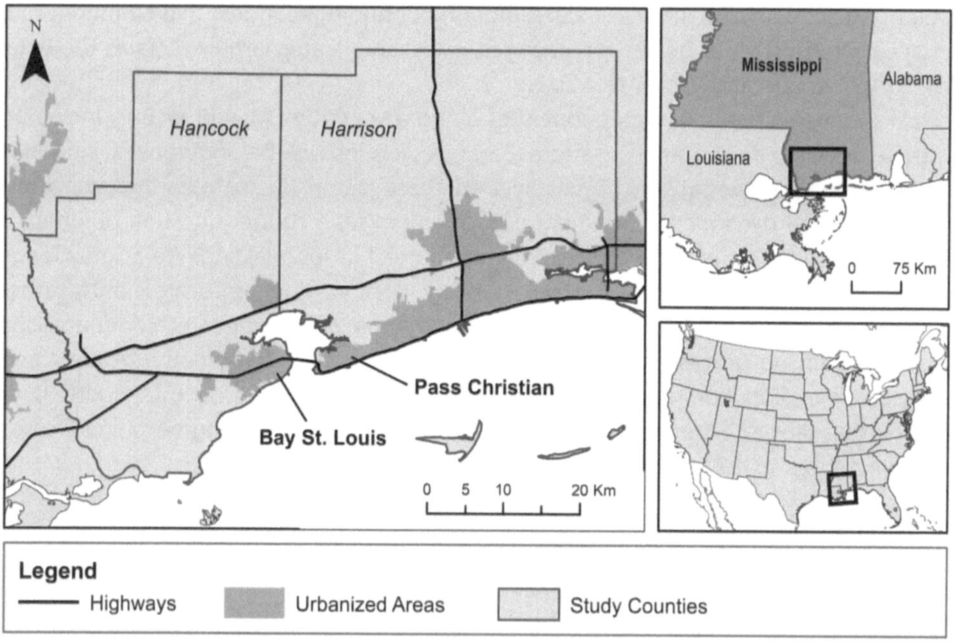

Figure 2. Bay St. Louis and Pass Christian, Mississippi Map.

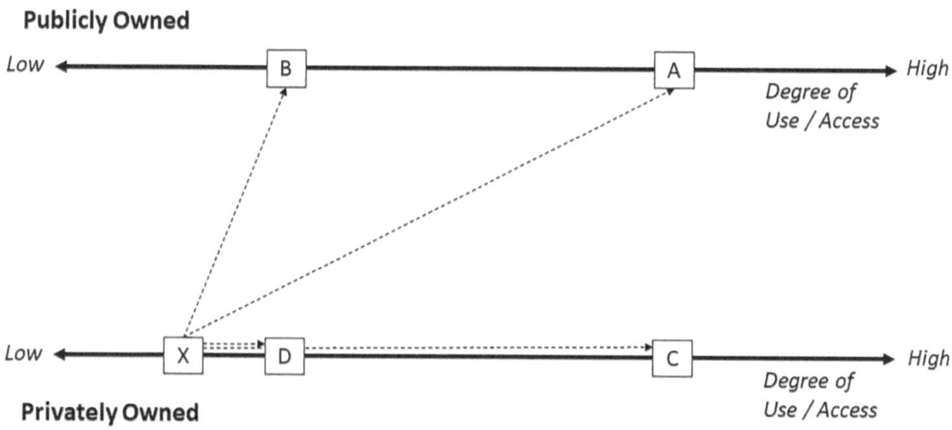

Figure 3. Communalism framework showing a privately-owned space before a disaster (X) becoming more public in terms of ownership (A, B) and/or degree of access (C, D) through post-disaster recovery.

communalism as observed in our fieldwork, showing the roles of two interrelated variables: property ownership and public access.

During post-disaster recovery, a variety of circumstances may induce private-to-public ownership change, such as buyouts. Yet not all post-disaster ownership changes transfer land from private to public domain. Improved building standards, changes in land zoning/permitting, and rising insurance prices may influence homeowners to sell single-family homes to private investors who then subdivide the lots into multi-family housing, thus distributing costs among more people. Although our findings suggest that communalism can occur irrespective of ownership changes, ownership status largely dictates the legal basis for public use and access to a space.

Unlike ownership, public access operates along a spectrum. Myriad factors, including land use policies, development patterns, and perceptions of belongingness, influence the degree to which a space is collectively used; these factors fluctuate during post-disaster recovery. The dynamic environment of recovery raises a further question of whether the observed patterns of communalism are ephemeral or long-lasting. We acknowledge these transitions in ownership and access occur over varying timescales; thus, rather than focus on time, we characterise the phenomenon of post-disaster communalism, describe its driving processes, and discuss its implications for community recovery and resilience. To explore communalism, we describe four cases (A, B, C, and D; Figure 3), or variants, that illustrate how differences in ownership status and degree of publicness spark differences in the theme of access.

4.1 Floodplain buyouts in Lexington, Kentucky

Analysis of survey responses and interview transcripts revealed access to and maintenance of public buyout open space as prominent themes in the neighbourhoods of Port Royal (Case A) and The Lane (Case B). The Lane, a private drive as indicated by signage, lies adjacent to the Port Royal neighbourhood and experiences flooding from the same stream, Wolf Run (Figure 1). A gate and brick fence restrict the public from entering the buyout property on The Lane. Conversely, the open space created through buyouts in Port Royal lacks any physical barriers, enabling unfettered public use of the land.

Residents of Port Royal repeatedly emphasised the presence of physical barriers on The Lane's buyout land; the gate restricts who can enter the public open space. One Port Royal resident commented via survey:

> One property that the city purchased for over $300,000 is not accessible to public. Is treated like private property by those living on THE LANE- Have a gate preventing anyone using the property as a shortcut. I feel I paid for it & should be able to … use it. I do understand that opening it to public increases security risk/vandalism etc. – mixed feelings. (Joanne; Zavar, 2014, p. 82)

By referencing that she paid for it (through taxes used to acquire the land), Joanne asserts ownership of the public space and argues that the gate prevents her from using the space. By referencing the buyout price of over $300,000, Joanne alludes to the socio-economic disparities between Port Royal and The Lane. Comparatively, the value of buyout properties in Port Royal averaged approximately $111,000 (Fayette County Property Valuation Administrator, 2011). The way Joanne emphasises the street name in her survey further suggests The Lane's elite status. Other Port Royal residents echoed similar sentiments. Gail expressed that '[p]art of [the] land on "The Lane" is not available to other residents. Land at end of our street is used by public. My husband goes down daily & gathers trash from area' (Zavar, 2014, p. 83). Gail also emphasises The Lane with quotations, reiterating street's prestigious status, and voices frustration at being excluded from the The Lane's public space while the open space in her neighbourhood remains publicly accessible.

Despite differences in physical barriers and public access to the open spaces, homeowners in both neighbourhoods indicated they contribute to the maintenance of the buyout properties, suggesting these buyout properties function as public-private hybrids (Nissen, 2008). Like Kayden (2000)'s POPS, legally the land is open to public access but is maintained by private entities. Federal cost-share grants mandate that acquired properties transfer to municipal ownership and remain open space in perpetuity (Zavar, 2015) yet private maintenance of the land is evidence of neoliberal processes that generate public-private hybrids. Gail explained that her husband cleans litter from the open space in Port Royal. Much of the litter is likely the result of public use, since the creek and adjacent open space serve as a boundary between the neighbourhood, a commercial district, and multi-family housing. Field observations and interviews suggest there is greater foot traffic along Port Royal's open space compared to The Lane's, where the brick fence seals out the public.

A homeowner on The Lane also contributes to the care of the buyout open space by paying for tree maintenance and mowing services. The homeowner viewed the maintenance provided by LFUCG as inadequate.

> Because the city's contract mowers did an infrequent & very poor job, I have been paying to maintain this property myself for more than two years. The city is aware of this. I have also paid individually for preventive treatment for the ash trees as the city SAID they would not do so. (Jamie; Zavar, 2014, p. 85)

Since LFUCG's maintenance of the open space does not meet this homeowner's standard, Jamie pays for manicured lawns and trees on the public buyout land. Their ability to incur this cost further indicates The Lane's affluence. Moreover, a private landowner paying for maintenance of the open space tacitly reinforces its aura of exclusiveness (Figure 4).

Figure 4. Street sign displaying private drive status (Author, 2013).

Both buyout properties are municipally-owned public spaces, yet residents perceive the open space in Port Royal as more public than the open space along The Lane (Figure 5). The lack of physical barriers in Port Royal enables public usage of the open space, resulting in litter. In contrast, residents perceive The Lane's open space as private given the gate, the private street designation, and the affluence of nearby residents. In referencing these cases to our post-disaster communalism framework (Figure 3), we observe that both properties transitioned to public ownership, but Port Royal (Case A) experiences a higher degree of public use and access than does The Lane (Case B).

4.2. Post-Katrina recovery in coastal Mississippi

Long-term recovery following Hurricane Katrina in coastal Mississippi resulted in increased public access to spaces that remained privately-owned. Communalism was evident along the beachfront where Katrina's destructive storm surge levelled many structures; new coastal views and improved access to beach amenities emerged post-disaster. We

Figure 5. (a) The Lane's open space (Author, 2013); (b) Port Royal's open space (Author, 2013).

observed communalism on beachfront parcels in Bay St. Louis where property owners opted not to rebuild after Hurricane Katrina (Case C). Whereas in Pass Christian, beachfront parcels redeveloped at higher densities than pre-disaster also displayed communalism (Case D).

Along sections of the beach in Bay St. Louis, vacant parcels largely outnumbered properties with rebuilt homes as of 2013 (eight years post-Katrina). Photovoice participants documented numerous locations where pre-disaster landmarks, including private single-family residences, never returned. Remnants of structures marked some parcels, but most remained vacant, either overtaken by vegetation or maintained by local lawncare companies. Two photovoice participants revealed that, while these private lots may no longer contain private homes, they are utilised in other distinct ways: public parking and public enjoyment of coastal views.

Observations and resident interviews suggest many local families park on the vacant private properties along Beach Boulevard across from the public beach (Figure 6). One participant, Kimberly, noted: 'It's interesting how a lot of this land has almost become public use. Everybody just parks in it to get to the beach.' Prior to Katrina, beachfront homeowners did not tolerate this sort of public appropriation—the threat they might call the police to tow a parked vehicle deterred such behaviour. Post-disaster, this void of surveillance increased public use of the private properties.

Several of the beachfront lots in this part of Bay St. Louis contained chairs, swings, and/or benches atop the remnant foundations (Figure 7). Despite the prohibitive cost of rebuilding and insuring a dwelling this close to the Gulf, many property owners maintained ownership to enjoy the beachfront amenities, namely gulf views and breezes (to combat the summer heat). Moreover, while property owners placed these swings and seats, the lack of fences or 'private property' signs invited broader public use. Based solely on the landscape, an observer unaware of the land tenure history would be unable to identify these as private properties. Hence, the coastal views and breezes once reserved for private property owners are now public commodities. It remains to

Figure 6. Vehicles parked on private property near Bay St. Louis (Schumann, 2015, p. 90).

Figure 7. Beachfront lot in Bay St. Louis used for public seating and gulf views (Schumann, 2015, p. 90).

be seen whether these spaces remain in public use over the longer term. Based on the communalism framework (Figure 3), Bay St. Louis (Case C) represents increased publicness and access to the beachfront amenities while remaining privately-owned.

Not all examples of post-Katrina communalism centred on open or vacant space. Case D in Pass Christian focuses on a common structure built after Katrina along the Mississippi Coast: condominiums. They, along with the single-family homes of affluent residents, are among the few structures to reemerge on the immediate coastline. The structure of the hazard risk insurance system in the United States strongly influences this redevelopment pattern.

Following a 1973 provision mandating National Flood Insurance Program (NFIP) flood insurance for homes purchased with federally-backed mortgages (Kousky, Lingle, & Shabman, 2017), low to moderate-income coastal homeowners began carrying the requisite flood insurance, while individuals purchasing homes outright with cash remained exempt. After Katrina, wealthier homeowners typically rebuilt first, unencumbered by a mortgage and the requirement to procure flood insurance. Meanwhile, homeowners with mortgages on decimated structures faced hurdles to reconstruction. In 2009, new NFIP Flood Insurance Rate Maps for the Mississippi Coast raised base flood elevations and increased premiums for new and rebuilt homes. Subsequently, the Biggert-Waters Flood Reform Act (2012) was passed to make the NFIP actuarially sound; the legislation raised artificially low flood insurance premiums to risk-based rates overnight.[1] Study participants with mortgages on single-family homes reported over 1,000 percent increases in their insurance rates.

> The closer you get to the beach, [...] the people aren't there. And, shortly after, people knew they wouldn't be able to [come back] – contractors knew they wasn't gonna [*sic*] be able to afford this 'cause, you know, wind and flood [insurance] ... And, sure enough, that's what's happened [...] a lot of people just can't afford it. When your insurance is higher than your mortgage – you just can't handle that (Royce).

> I did get a quote–the actuarially sound number from our flood insurance. I'm paying [...] about $500.00 right now for an accident out ... at which you can get $250,000 coverage for the

house, and I think $100,000 in contents. My actuarially sound [rate] is a little over $21,000 a year. [...] Nobody can afford that. (Elaine).

In instances where condos were built (Figure 8), developers combined multiple parcels and rebuilt at higher densities for profitability. This strategy spreads the risk burden (i.e. higher insurance premiums) among more housing units, making beachfront property ownership more affordable for prospective buyers. Yet the higher density per parcel also means greater numbers of people living in a higher-risk landscape. Interviewees expressed varying opinions regarding the trend toward higher-density housing post-hurricane and its effect: more and different people now had access to beachfront amenities. Although still privately owned, the condominiums expanded access to gulf views and beaches across a wider socioeconomic spectrum, including tourists, compared to pre-Katrina. Study participants expressed mixed concerns.

Buying the land – condo, condo, condo, condo. And that's the only thing I don't like is, because [before the storm] it was more of a home feeling. And then, you start putting condos in, and now, you're on the beach, and [it feels like] Florida, you know? [...] more tourists and business and not good Mississippi folk (Royce).

I like the changes. The old antebellum homes [on the beach] were nice to look at, but I mean, they didn't really help anybody. Certain families owned them. Nobody else could afford to live around there on the beach. If you didn't have money, you were never going to be there. Now, those homes are gone [...] you've got places like this, large condos [...] 25 or 30 units. Now

Figure 8. (a) Rebuilt condominium complex in pass Christian (Schumann); (b) View across highway toward beach (Schumann, 2015, p. 93).

> you got 30 different families that now can have a beachfront view. [...] It gives a lot of people an opportunity to live on the beach [...] Quite a few friends of mine rent places like this [...] Most of them are rental properties (Justin; Schumann, 2015, p. 92).

Royce expressed concern over the Mississippi coastline losing its sense of place through excessive condominium development, an issue identified in other post-disaster communities (e.g. Chamlee-Wright & Storr, 2009; Erikson, 1976). Further troubling to Royce was the increased number of outsiders replacing 'good Mississippi folk' who resided along the beachfront before Katrina. Yet, not all study participants viewed the changing demographics negatively. As Justin indicated, families of lower socio-economic status were now able to live beachfront in the condos, whereas prior to Hurricane Katrina, the land was owned by a small number of wealthy families. Despite the beachfront property remaining in private ownership, Justin valued the increased access to coastal amenities for residents and the public at-large.

Nearly 15 years post-Katrina, Mississippi's coastline remains a checkerboard of rebuilt and vacant parcels. Where rebuilding is concentrated, more residents and visitors can access the beach and its amenities through property ownership, rental, or use of public facilities. Based on the communalism framework (Figure 3), parcels in Pass Christian where condominiums replaced single-family homes post-Katrina (Case D) exhibit increased usage and access by a wider cross-section of the public. The land remains privately-owned, but the ownership and development patterns afford beachfront access to more people through the condominium-style development.

5. The mechanisms of communalism

While a disaster event initiated the observed changes in land use and/or ownership, we discuss five different mechanisms drawn from existing geographic scholarship that shape residents' evolving perceptions about post-disaster land use in these communal spaces. The mechanisms include aesthetics, security, pricing, amenities, and designation. Using the case studies, we discuss how these mechanisms contribute to the perceived publicness of formerly exclusive spaces. These mechanisms operate collectively in compressed time during post-disaster recovery.

5.1. Aesthetic mechanism

The aesthetic mechanism focuses on the visual appearance of a space—how its design, layout, and level of maintenance imply who may use it. Scholars have highlighted how methods for landscape design and management communicate notions about a space's exclusivity or belongingness for select groups (Duncan & Duncan, 2001). In the Lexington case, the appearance and level of care in The Lane's open space differs dramatically from Port Royal's. Through multiple visits and prolonged field study, litter was always visible in the creek along the open space in Port Royal. The grass in the accessible open space was often taller than in neighbouring yards of private homes, lending an unkempt appearance. By comparison, The Lane's manicured appearance, like a private garden tucked behind the ivy-covered fence, suggests exclusivity (Brown, 1987). The space's frequently mowed grass and neatly trimmed trees—the handiwork of professional landscapers paid for by private homeowners—reinforces the landscape's private feel. Similarly, in Bay St. Louis, the beachfront properties were regularly mowed and maintained by either the private homeowners or hired lawncare companies; however, the parcels lacked physical barriers. Owners placed benches, chairs, and swings—features common in public parks—that invited the public to also use the space. These factors generated greater access to these private spaces than was experienced prior to Hurricane Katrina. Rather than displaying territoriality, aesthetics communicated inclusion and effectively rendered these private lots hybrid spaces. Similarly, in Pass Christian, the condos depict inclusive aesthetics despite their private ownership (Figure 7). Oriented for catching gulf breezes, these units face outward toward the beach. The lack of gates or fences is notable; one may readily access any front door in the complex. Other condominium complexes reflect varying degrees of private aesthetization by situating units to face inward courtyards or using coded gate entries to prevent encroachment by outsiders.

5.2 Security mechanism

Closely tied to aesthetics is the second mechanism, security. The previous examples show that physical barriers (or lack thereof) act not only as design elements, but also to control ingress and egress. Once installed, these barriers prevent the general public from gaining access to the space. The entry gates to The Lane reduced usage of the municipally-owned space. These gates and similar security measures are identified in the privatisation literature as prominent landscape features used to control public access and use (e.g. Davis, 1990; Mitchell, 2010). By contrast, a lack of gates in Bay St. Louis invited the public to enjoy the beachfront land despite its private ownership. Meanwhile, the ungated, outward-facing condos in Pass Christian suggest anyone may lease them or enter the premises.

5.3 Economic mechanism

The economic mechanism involves the actual or perceived land value. Beyond contributing to perceptions about a space's publicness or exclusivity, economic tools like tax valuations and rental rates impose practical barriers to land ownership and usage. For example, the property values along The Lane were higher than those in Port Royal. Study participants related these economic differences when voicing their frustrations over their level of access to each property. The higher cost of the property along The Lane made the street, and the publicly-owned land within, seem more exclusive compared to the less expensive properties in Port Royal. Additionally, the economic mechanism encompasses chronic expenses associated with living in a high-risk landscape. As observed in Pass Christian, flood insurance is unaffordable for many homeowners. Facing these economic constraints, property owners impacted by Hurricane Katrina were left with two strategies: either do not rebuild or subdivide a parcel into condominiums where the cost burden, and risk, is shared. Although the first strategy implies decreased intensity of land use by current owners, the second implies both increased use by a wider range of individuals and potential loss of ownership. Both could yield spaces perceived as more public.

5.4 Amenity mechanism

Amenities comprises an array of features that provide recreational enjoyment or enhance residents' quality of life (Bin et al., 2008). Amenities include proximity to beaches, gulf breezes, water views, municipal piers, among others. In Bay St. Louis, these amenities became more accessible to the public following Hurricane Katrina as the soaring insurance costs contributed to beachfront properties remaining vacant. Unable to rebuild their homes due to unfavourable economics, landowners installed additional amenities (e.g. swings and benches) inviting the public to enjoy the Gulf view and breezes. In Pass Christian, subdivision of property and risk-spreading through denser condominium development enabled the construction of community pools and tennis courts, affording additional amenities to those able to pay the requisite rent. In Port Royal, the neighbourhood association developed a niche amenity, a labyrinth, as a site of reflection for public use. Although used infrequently, the labyrinth is listed on a national registry, and healthcare groups have utilised it to assist in patient recovery (Zavar, 2016). These examples demonstrate how, despite constraints imposed by economics and/or security, amenities invite the public to engage more extensively with the land post-disaster, thus increasing access to and use of these once private spaces.

5.5 Designation mechanism

Designation is a type of place making that suggests or prescribes the set of possible functions and meanings for a space. Designation can include giving names to previously un-named places, (re)branding spaces, or codifying appropriate uses of a landscape through rules, regulations, and signage. Although both aesthetics and designation mechanisms influence sense of place, aesthetics focuses merely on the implications of appearance for use and ownership, while designation points to the statuses, monikers, and political structures responsible for inscribing those norms of use and ownership. For example, The Lane's street sign has formal distinction as a private street, represented by the letters 'PVT.' The signage reinforces a sense of privilege, which is sanctioned by LFUCG, thus contributing to the private feel of the publicly-owned open space. Coupled with the fences, the signage demonstrates territoriality over the manicured plot through both demarcation and defense (Brown, 1987). Conversely, it is the lack of signage and barriers along vacant beachfront properties in Bay St. Louis that encourages public use of the benches. Without formal designation, such as 'private property' markers or 'no parking' signs, the perceived exclusivity of the private land is diminished and public use increases. In the above examples, signage alone does not shape perceptions about ownership privileges or rights of use; rather the amenities, economic conditions, aesthetics, and security mechanisms act in tandem to facilitate or hinder usage. With an understanding of how the five mechanisms of communalism are entangled, we next consider the implications of post-disaster communalism.

5.6 Implications of communalism

The cases of (A) Port Royal, (B) The Lane, (C) Bay St. Louis, and (D) Pass Christian mirror private-public hybridisation found in previous scholarship (Kayden, 2000; Nissen, 2008), yet they pose different implications for public space and for community resilience to future disasters. At best, post-disaster communalism promotes increased interaction among community members, thus evoking the ideals of public space (e.g. Davis, 1990) and positively impacting resilience by strengthening social ties and place attachments. Yet, in each of the four cases, distinct tradeoffs exist between promoting ideals of public space and bolstering resilience.

The public-private hybrid examples in Lexington, Kentucky, demonstrate that communalism does not inherently provide increased access to public space. While buyouts require private to public ownership change, access to the public space varies site to site, as with POPs (Kayden, 2000). For example, some municipalities rent the buyout land to adjacent homeowners for nominal fees. In return, the neighbour maintains the buyout land, thus reducing the financial burden on municipalities (Zavar & Hagelman, 2016). This arrangement, best described as publicly-owned private space, most aligns with the experience along The Lane in Lexington. Alternatively, communities have transformed buyout properties into community parks, gardens, and other publicly accessible spaces (Zavar & Hagelman, 2016). Closely resembling the Port Royal experience, these land uses invite greater public use and, by extension, more public interaction than does the rented land.

The variability in access to buyout land observed in Lexington, Kentucky, offers implications for building long-term community resilience and questions the notion that buyout land ought to serve the greater public interest. The creation and/or protection of open space may be an important mitigation tool to reduce community-wide flood losses and build resilience to future floods (Brody & Highfield, 2013), yet how buyout land is perceived and managed impacts community support for these mitigation programmes (Zavar, 2015). Despite the increasing popularity of residential buyouts to reduce occupation of high-risk areas (Greer & Binder, 2017), permanent relocation facilitated through buyout programmes can lead to negative consequences for participants including loss of social networks and sense of place (Binder & Greer, 2016). These programmes can be contentious for impacted residents, governmental agencies, and other stakeholders. Adjacent homeowners that were not selected to participate in the buyout —yet are still impacted by flooding—can feel unprotected; hence, community engagement and inclusive practices are essential in reducing these tensions (Binder & Greer, 2016). To increase community support for residential buyouts, open space uses and management practices must meet the needs of community members inside and outside of buyout areas. Furthermore, buyouts offer the opportunity for green infrastructure, like wetland restoration, to both improve ecosystem services and reduce future flood-risk across the city (Harter, 2007; Zavar, 2015). The exclusionary practices observed in Lexington, Kentucky, could inflame tensions over buyout properties, potentially undermining support for future mitigation or open space projects, including green infrastructure, which could provide city-wide benefits. In this way, communal land use practices could inadvertently thwart long-term efforts toward disaster resilience.

In Mississippi, although no formal buyout programme ensued, post-Katrina adjustments to the flood insurance system rendered a landscape replete with vacant parcels, akin to a buyout. New NFIP policies and flood maps staved off most redevelopment, thereby reducing lives and property at risk while preserving the empty coastline as a reminder of Katrina's devastation (Colten & Giancarlo, 2011). Although heightened social memory and fewer residences in harm's way bode well for long-term disaster resilience, individual parcels exist, like those in Pass Christian, where the number and vulnerability of residents has increased post-disaster. New condo residents tend to be renters rather than owners, making them more sensitive to hazards given their limited power to repair, harden, or insure the structure (Morrow, 1999). Relative to homeowners, renters tend to have lower socioeconomic status and often lack the place attachments and strong social ties necessary for expedient disaster response and recovery (Burby, Steinberg, & Basolo, 2003). Furthermore, renters often struggle to find replacement housing post-disaster because of the slower repair of rental units and their unaffordability (Rufat, Tate, Burton, & Maroof, 2015). If these residents are also new to the region, they may be unfamiliar with endemic risks and lack knowledge on preparedness.

Therefore, the land use patterns emergent in Hurricane Katrina's aftermath, observed in Bay St. Louis and Pass Christians, carry a double-edge when considering their consequences for Mississippi's resilience to future disasters. If history is a guide (Colten & Giancarlo, 2011; Cutter et al., 2014), the empty beachfront parcels will be redeveloped. However, when redevelopment occurs, will a system exist for the public acquisition of these lands? Or will lands be resold to developers to feed the recovery growth machine (Pais & Elliott, 2008), thus repopulating the coast at ever-higher densities and erasing Katrina's last vestiges?

6. Conclusions

Our case studies identify the emergence of quasi-public spaces after a disaster, which we term post-disaster communalism. Formerly private land transitioned to greater public use, and in some instances, ownership. Consequently, perceptions about public access to the land also changed. Although communalism does not occur in all disaster impacted areas, the potential exists for environmental hazards like flooding and hurricanes to blur the public-private divide. As climate change contributes to more extreme weather patterns, it is likely that coastal and riverine cities will increasingly pursue buyouts and organised retreat strategies, thus increasing the potential for communal land use practices.

The present study contributes to scholarship by considering communal land ownership and management as both a post-disaster outcome impacting public space and as a phenomenon that shapes future community resilience. Future research should compare the ways in which communalism impacts disaster recovery and resilience following other types of hazards (e.g. wildfires, tornadoes, earthquakes) and across a range of socio-political landscapes. Additional work is needed to understand the timeframe of communalism: are these processes temporary shifts or longer-term patterns within the macroscale tendency toward neoliberalism? Finally, since buyout policies and insurance market conditions contributed to communal land use, future research should identify other political-economic forces capable of interacting to induce post-disaster communalism.

Note

1. For discussions on flood insurance history, pricing, and reform see Kousky et al. (2017) and Zhao, Kunreuther, and Czajkowski (2015).

Disclosure statement

No potential conflict of interest was reported by the authors.

Funding

This work was supported in part by National Science Foundation [grant number 1301830].

ORCID

Elyse M. Zavar http://orcid.org/0000-0001-5017-8411
Ronald L. Schumann http://orcid.org/0000-0002-2293-6331

References

Bin, O., Crawford, T., Kruse, J., & Landry, C. (2008). Viewscapes and flood hazard: Coastal housing market to amenities and risk. *Land Economics, 84*(3), 434–448.

Binder, S. B., & Greer, A. (2016). The devil is in the details: Linking home buyout policy, practice, and experience after Hurricane Sandy. *Politics and Governance, 4*(4), 97–106.

Bolitzer, B., & Netusil, N. R. (2000). The impact of open spaces on property values in Portland, Oregon. *Journal of Environmental Management, 59*, 185–193.

Brody, S., & Highfield, W. (2013). Open space protection and flood mitigation: A national study. *Land Use Policy, 32*, 89–95.

Brower, S. (1980). Territory in urban settings. In I. Altman, A. Rapoport, & J. Wohlwill (Eds.), *Human behavior and environment* (pp. 179–207). Boston, MA: Springer.

Brown, B. (1987). Territoriality. In D. Stokols & I. Altman (Eds.), *Handbook of environmental psychology* (Vol. 1, pp. 505–533). New York: John Wiley & Sons.

Brown, B., & Perkins, D. (1992). Disruptions in place attachment. In I. Altman, & S. Low (Eds.), *Place attachment* (pp. 279–304). Boston, MA: Springer.

Burby, R., Steinberg, L., & Basolo, V. (2003). The tenure trap: The vulnerability of renters to joint natural and technological disasters. *Urban Affairs Review, 39*(1), 32–58.

Chamlee-Wright, E., & Storr, V. (2009). "There's no place like New Orleans": sense of place and community recovery in the ninth ward after Hurricane Katrina. *Journal of Urban Affairs, 31*(5), 615–634.

Charmaz, K. (2006). *Constructing grounded theory: A practical guide through Qualitative analysis.* Thousand Oaks, CA: Sage.

Colten, C., & Giancarlo, A. (2011). Losing resilience on the gulf Coast: Hurricanes and social memory. *Environment: Science and Policy for Sustainable Development, 53*(4), 6–19.

Cutter, S., Emrich, C., Mitchell, J., Piegorsch, W., Smith, M., & Weber, L. (2014). *Hurricane Katrina and the forgotten coast of Mississippi.* Cambridge: Cambridge University Press.

Davis, M. (1990). *City of quartz excavating the future of Los Angeles.* London: Verso.

Dillman, D. (1978). *Mail and telephone surveys: The total design method.* New York, NY: Wiley & Sons.

Duncan, J., & Duncan, N. (2001). The aestheticization of the politics of landscape preservation. *Annals of the Association of American Geographers, 91*(2), 387–409.

Erikson, K. (1976). *Everything in its path: Destruction of community in the buffalo creek flood.* New York, NY: Simon and Schuster.

Fairclough, N. (1995). *Critical discourse analysis.* Boston, MA: Addison Wesley.

Fayette County Property Valuation Administrator. (2011). Web access to property records. Retrieved from http://qpublic9.qpublic.net/ky_fayette_search2.php

FEMA. (1998). *Property acquisition handbook for communities: Phase IV open space management.* Washington DC: Federal Emergency Management Agency.

Geipel, R. (1991). *Long-term consequences of disaster: The reconstruction of Friuli, Italy, in its international context 1976–1988.* New York, NY: Springer-Verlag.

Godschalk, D., Beatley, T., Berke, P., Brower, D., & Kaiser, E. (1999). *Natural hazard mitigation recasting disaster policy and planning.* Washington DC: Island Press.

Greer, A., & Binder, S. B. (2017). A historical assessment of home buyout policy: Are we learning or just failing? *Housing Policy Debate, 27*(3), 372–392.

Haas, E., Kates, R., & Bowden, M. (1977). *Reconstruction following disaster.* Cambridge, MA: MIT Press.

Hagelman III, R., Connolly, M., Zavar, E., & Dahal, K. (2012). Disaster reconstruction and business geography following the 2007 Greensburg, Kansas Tornado. *Environmental Hazards, 11*(4), 283–302.
Harter, J. (2007). *Riparian restoration: An option for voluntary buyout lands in New Braunfels, TX (doctoral dissertation)*. San Marcos: Texas State University.
Harvey, D. (2005). *A brief history of neoliberalsim*. Oxford: Oxford University Press.
Kayden, J., & the New York City Department of City Planning, & the Municipal Art Society of New York (2000). *Privately owned public space: The New York city experience*. New York, NY: Wiley & Sons.
Klein, N. (2007). *The shock doctrine: The rise of disaster capitalism*. New York: Metropolitan Books.
Knabb, R., Rhome, J., & Brown, D. (2005). *Hurricane Katrina*. Miami, FL: National Hurricane Center/ Tropical Prediction Center.
Kousky, C., Lingle, B., & Shabman, L. (2017). The pricing of flood insurance. *Journal of Extreme Events, 4* (2), 1750001.
McCann, E. (1997). Where do you draw the line? Landscape, texts and the politics of planning. *Environment and Planning D: Society and Space, 15*, 641–661.
Mitchell, D. (2010). The end of public space? People's park, definitions of the public, and democracy. In A. Orum, & Z. Neal (Eds.), *Common ground? Readings and reflections on public space* (pp. 83–99). New York, NY: Routledge.
Montz, B. (1987). Floodplain delineation and housing submarkets: Two case studies. *The Professional Geographer, 39*(1), 59–61.
Morrow, B. (1999). Identifying and mapping community vulnerability. *Disasters, 23*(1), 1–18.
Németh, J. (2009). Defining a public: Managing privately owned public space. *Urban Studies, 46*, 2463–2490.
Nissen, S. (2008). Urban transformation from public and private space to spaces of hybrid character. *Czech Sociological Review, 44*(6), 1129–1149.
NVivo. (2012). *Qualitative data analysis (Version 10) [software]*. QSR International Pty.
Pais, J., & Elliott, J. (2008). Places as recovery machines: Vulnerability and neighborhood change after major hurricanes. *Social Forces, 86*(4), 1415–1453.
Rosen, C. (1986). *The limits of power: Great fires and the process of city growth in America*. Cambridge: Cambridge University Press.
Rubin, H., & Rubin, I. (1995). *Qualitative Interviewing: The art of hearing data* (2nd ed). Thousand Oaks, CA: Sage Publications.
Rufat, S., Tate, E., Burton, C., & Maroof, A. (2015). Social vulnerability to floods: Review of case studies and implications for measurement. *International Journal of Disaster Risk Reduction, 14*, 470–486.
Schumann, R. (2015). *The meaning of place recovery on the Mississippi gulf Coast (doctoral dissertation)*. Columbia: University of South Carolina.
Schumann, R., Binder, S. B., & Greer, A. (2019). Unseen potential: Photovoice methods in hazard and disaster science. *GeoJournal, 84*(1), 273–289.
Sibley, D. (1995). *Geographies of exclusion*. London: Routledge.
Trudeau, D. (2006). Politics of belonging in the construction of landscapes: Place-making, boundary drawing and exclusion. *Cultural Geographies, 13*, 421–443.
Trujillo-Pagan, N. (2012). Neoliberal disasters and racialisation: The case of post-Katrina Latino labour. *Race & Class, 53*(4), 54–66.
Zavar, E. (2014). *The value of open space: A geographic case study of floodplain buyouts in Lexington, Kentucky (doctoral dissertation)*. San Marcos: Texas State University.
Zavar, E. (2015). Residential perspectives: The value of floodplain-buyout open space. *Geographical Review, 105*(1), 78–95.
Zavar, E. (2016). The role of magnetic agents in land use decision-making and neighborhood initiatives. *Land Use Policy, 56*, 38–46.
Zavar, E., & Hagelman, R. (2016). Land use change on U.S. Floodplain buyout sites, 1990-2000. *Disaster Prevention and Management, 25*(3), 360–374.
Zavar, E., Hagelman, R., Lavy, B., & Prince, B. (2017). Land use change at temporary group-housing sites in post-Katrina Louisiana. *Natural Hazards Review, 18*(3), 4017003.
Zhao, W., Kunreuther, H., & Czajkowski, J. (2015). Affordability of the national flood insurance program: Application to Charleston County, South Carolina. *Natural Hazards Review, 17*(1), 04015020.
Zukin, S. (1995). *The Culture of cities*. Cambridge: Blackwell.

Prospects for disaster management in China and the role of insurance

Xian Xu

ABSTRACT
Natural disasters have serious negative consequences for China and it is necessary to build an effective and efficient disaster relief system. This paper aims to provide suggestions for how to restructure and optimise China's disaster relief system. This paper first discusses the four main channels through which relief funds are currently distributed in China while also examining the relative share of relief funds directed through each channel. Then, the advantages and disadvantages of these relief channels are compared. Finally, suggestions for how China can reduce the negative economic and social impacts of natural disasters by restructuring and optimising its current disaster relief system are provided. The paper presents several main findings. Currently, government-channelled funds are the most important source of disaster relief in China. However, the actual ratio of relief funds from the government to the total amount of losses and the ratio of relief funds from the four channels added together are both very low. This paper argues that the role of commercial insurance in disaster relief is far from sufficient. Importantly, suggestions are also provided on how to restructure the system and on the relative role that each of these channels should play in China's disaster relief system.

1. Introduction

China is one of the most severely affected countries by natural disasters, with a variety of such events occurring every year. Since the beginning of the twenty-first century, catastrophes have occurred frequently, including the Zhouqu mudslides, the Wenchuan and Yushu earthquakes, annual summer typhoons in coastal areas, floods in the eastern region, and droughts in the western region. These disasters have resulted in large economic losses for the nation and have created substantial pressure for state fiscal disaster relief. Hence, research on disaster economics and suggestions for improvement of the disaster relief system are of great significance to China's economic development (Liu, 2009).

Because disasters are becoming more frequent, and the world's population has become more concentrated and the ecological environment has deteriorated with rapid economic development, disasters are causing greater economic losses than ever before. In 2016, the

Table 1. Economic growth and the losses from natural disasters in China.

Year	Total losses from natural disasters (in 100 million RMB)	Growth rate of disaster losses (in percentages)	GDP	Natural disaster losses/GDP (in percentages)
2001	1942.2	−5.04	109,276.2	1.78
2002	1717	−11.60	120,480.4	1.43
2003	1884.2	9.74	136,576.3	1.38
2004	1602.3	−14.96	161,415.4	0.99
2005	2042.1	27.45	185,998.9	1.10
2006	2528.1	23.80	219,028.5	1.15
2007	2363	−6.53	270,844	0.87
2008	11,752.4	397.35	321,500.5	3.66
2009	2523.7	−78.53	348,498.5	0.72
2010	5339.9	111.59	411,265.2	1.30
2011	3096.4	−42.01	484,753.2	0.64
2012	4185.5	35.17	539,116.5	0.78
2013	5808.4	38.77	590,422.4	0.98
2014	3373.8	−41.92	644,791.1	0.52
2015	2704.1	−19.85	686,449.6	0.39
2016	5032.9	86.12	740,598.7	0.68
Average	3618.5	31.85	373,188.5	1.15

Note: The data include disaster relief funds from all departments of the central government, not only from the Civil Affairs Ministry.
Source: 'China Statistical Yearbook' (2017), 'China Civil Affairs Statistical Yearbook' (2001–2017).

world's natural and man-made disasters caused more than 11,000 deaths and missing individuals as well as economic losses of nearly $175 billion (Swiss Re, 2017).

Table 1 shows the relationship between GDP and the direct economic losses caused by natural disasters in China from 2001 to 2016. During this period, the annual economic losses from natural disasters averaged 361.85 billion RMB.

The serious impact of natural disasters on the stable development of China's economy and society means that building an effective system for disaster relief is indispensable for maintaining social stability and economic development in the country.

2. The composition of China's current disaster relief system

China's current disaster relief system is mainly composed of four major entities: the government, insurance, social organisations, and victims. Insurance includes disaster insurance with government participation and commercial insurance with market operations. When a disaster occurs, the government dispenses funds for assistance in the affected areas and organises relevant agencies to support victims. At the same time, under the guidance of the government, social organisations such as charities also provide material and human resource assistance to the affected areas. Furthermore, insurance also participates in the process. Insurance claims are made, and insurance companies compensate victims for the portion covered by the insurance policy, but funding is not always available for catastrophic insurance from the government. After the disaster, victims' self-help and mutual assistance are still the main forms of disaster relief; the government organises reconstruction and other post-disaster services such as providing psychological comfort for those affected and education on disaster prevention and reduction (see Figure 1).

As a result, disaster relief can be divided into four channels. The most important channel is government agencies supported by national financing. The other channels are spontaneous non-government charitable institutions and national charitable

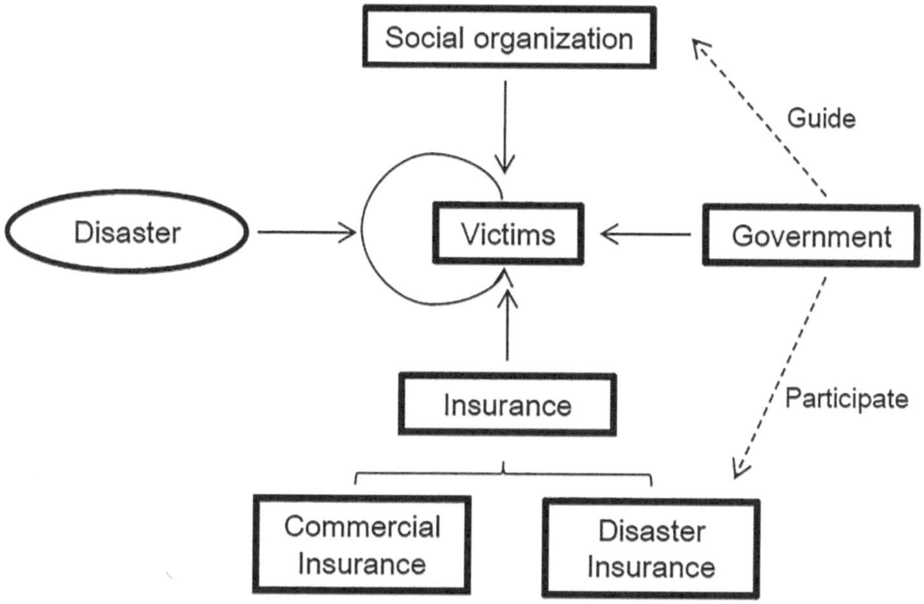

Figure 1. Main channels in China's current disaster relief system.

organisations, commercial insurance, and mutual assistance and self-help provided by the disaster-affected population.

2.1. Government assistance

Government assistance consists of reserve funds set aside in the government's budget to address a wide variety of natural disasters in China by, for example, providing financial and material assistance to the disaster-affected areas. Government assistance can be divided into central government and local government fiscal rescues. The central government's fiscal rescue includes civil affairs relief funds for victim relief on the one hand and funds for infrastructure reconstruction on the other hand. The natural disaster relief work by local governments at or above the county level is subordinated to the national economic and social development plan. Meanwhile, natural disaster relief funds and work expenses are also accounted for in local budgets (National Disaster Reduction Committee, Article 4).

The national comprehensive emergency coordination institution for natural disaster relief in China is the National Disaster Reduction Committee, which is responsible for organising and leading national natural disaster relief work and coordinating rescue activities for special major and major natural disasters (National Emergency Plan for Natural Disaster Relief, 2016). When the risk of a disaster increases, the meteorology, water resources and other departments report disaster warning information to the National Disaster Reduction Commission in a timely manner, and then the Commission engages in response measures to initiate early warnings. When the disaster risk turns into an actual disaster, different levels of national emergency response countermeasures are taken based on the degree of the risk and other influencing factors. After the disaster, the National Disaster Reduction Committee organises related departments and experts to conduct disaster assessments and carry out post-disaster relief procedures, such as life support, during

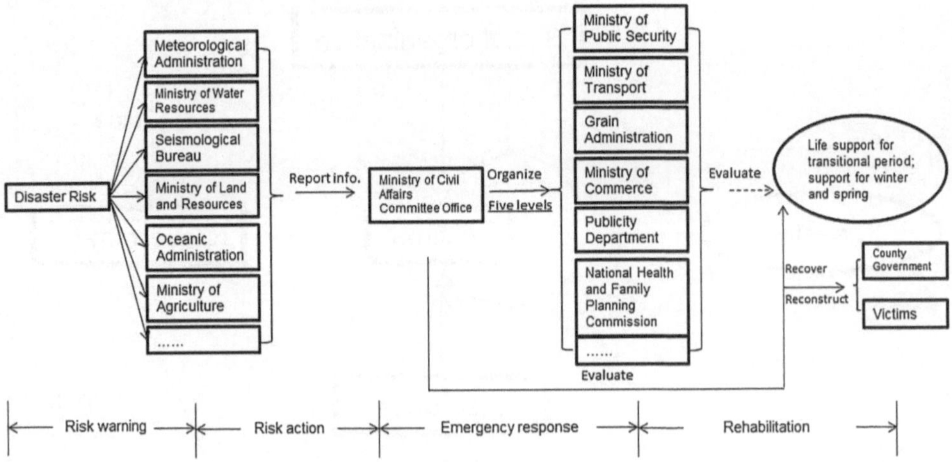

Figure 2. Government action during different disaster phases.

the transitional period and in winter and spring rescues. The reconstruction of damaged houses is mainly performed by the affected households themselves based on a series of preferential policies that are organised and implemented by the county-level government (see Figure 2).

As shown in Table 2, relief spending by the government increased yearly until 2012, when it reached its peak, but since then, relief spending has decreased. Despite the volatility of relief expenditures, only a very low proportion of the total losses are covered, ranging from 2.11% to 7.89% in the years surveyed.

2.2. Insurance payments

2.2.1. Catastrophe insurance

In China, catastrophe insurance is led by the government, although the market participates in its operation (Wei & Zhang, 2012). The government provides insurance compensation for people who have suffered from disasters through the establishment of catastrophe insurance funds and other methods, usually through enforcing basic disaster insurance, which reduces the financial burden on the government and guarantees the basic livelihood of the affected people.

Since 2014, the China Banking and Insurance Regulatory Commission[1] (CBIRC) has conducted pilot programmes for catastrophe insurance in Shenzhen, Ningbo and Yunnan through the model of 'government-led commercial insurance participation'. The government helps residents purchase catastrophe insurance from insurance companies and actively encourages commercial insurers to offer more personalised catastrophe insurance products to residents. In recent years, several provinces have established catastrophe insurance, as shown in Table 3.

China's experience with these pilot programmes has led to substantial progress in the establishment of the country's own catastrophe insurance system; presently, several companies include earthquake insurance products in common property insurance (Fan, 2016; He, 2017). However, as we can see from the Table 4, the catastrophe insurance premium

Table 2. Government assistance compared to natural disaster losses 2001–2016.

Year	Natural disaster total losses (100 million RMB)	Fiscal relief expenditures[a] (100 million RMB)	Relief ratio of total losses (in percent)
2001	1942.2	41.0	2.11
2002	1717.0	40.0	2.33
2003	1884.2	52.9	2.81
2004	1602.3	51.1	3.19
2005	2042.1	62.6	3.07
2006	2528.1	79.0	3.12
2007	2363.0	79.8	3.38
2008	11,752.4	609.8	5.19
2009	2523.7	199.2	7.89
2010	5339.9	237.2	4.44
2011	3096.4	231.7	7.48
2012	4185.5	272.0	6.50
2013	5808.4	240.9	4.15
2014	3373.8	98.73	2.93
2015	2704.1	94.72	3.50
2016	5032.9	79.1	1.57
Average	3618.5	154.36	3.98

Sources: 'China Social Statistical Yearbook' (2016), 'China Civil Affairs Statistical Yearbook' (2001–2016), 'China National Public Finance Expenditure Statement Report' (2011–2016).

[a]The appropriation from the government, which is used to provide assistance in disaster areas, such as emergency rescue, solatium for family members of the victims, transitional life assistance, restoration and reconstruction subsidies for collapsed or damaged houses, etc. (National Disaster Reduction Committee, 2016).

Table 3. Catastrophe insurance established in several provinces.

Province	Burden form of premium for catastrophe insurance	Type of catastrophe risks	Insurance companies participating[a]
Shenzhen	Municipal government	Several disasters, i.e. typhoon, flood, etc.	3
Zhejiang (Ningbo)	Municipal government	Several disasters.	6
Yunnan	Provincial, municipal and county financial commitment	Earthquake	45
Guangdong	Provincial and municipal finance jointly support the program[b]	Several disasters	3
Sichuan	Provincial finance, municipal and county finance, ordinary urban and rural residents.	Earthquake	41
Shanghai	Municipal Finance Bureau	Several disasters	3

Source: The local branches of CBIRC.

[a]The reinsurance companies are not included.

[b]Each pilot city owns a budget of 30 million yuan. If the premium is less than 30 million yuan, the premium shall be shared in a proportion of 3:1 between the provincial level and the municipality, and the portion of the premium exceeding 30 million yuan shall be borne by the municipality.

income only accounts for a very small fraction of the total property insurance premium income; the growth rate of the premium is quite fluctuant. As of August 31, 2018, there are only 23 pilot cities offering catastrophe insurance, accounting for only 3.5% of the total number of cities.[2]

The market supply remains less than the demand; for instance, Li, Zhang, and Guo (2015) have shown that there is an obvious gap and structural problems with the supply of flood insurance by comparing the risk level, which represents the demand, with the insured liabilities in different policies, which represents the supply.[3]

Table 4. China's catastrophe insurance premiums.

Year	Total premium (property insurance) (100 million yuan)	Growth rate of gross premium (property insurance)	Catastrophe insurance premium (100 million yuan)	Growth rate of premium (Catastrophe insurance)	Catastrophe insurance premium/ property insurance premium
2014	7203.38	16.0%	0.74	–	0.01%
2015	7994.97	11.0%	0.62	−16.8%	0.01%
2016	8724.50	9.1%	1.51	144.7%	0.02%
2017	9834.66	12.7%	2.53	68.1%	0.03%
2018 (January to September)	8003.40	−18.6%	2.41	−5.0%	0.03%

Source: China Banking and Insurance Regulatory Commission (CBIRC).

Moreover, the disaster programmes have many deficiencies, such as low participation rate by commercial insurance companies. As shown in Table 3, except for Yunnan and Sichuan,[4] there are only a few insurance companies that have participated in this programme, which may cause heavy burdens for the participating companies when a disaster happens, or even bankruptcy if there are not enough members with whom to share the risks.[5]

Here we can briefly summarise that the implementation scope and maturity are very low, and are still being explored and cultivated.

2.2.2. *Commercial insurance*

Commercial insurance compensates the insured for property damage and losses caused by various types of disasters and accidents through insurance payments. It is one of the most important channels for post-disaster relief, and it contributes to maintaining economic and social stability in most developed countries. Moreover, commercial insurance provides incentives for policyholders to actively engage in pre-disaster precautionary behaviours. The world's average insurance payout rate for disaster losses was approximately 36% in 2008 (Wang, 2008). In 2013, the economic losses from catastrophes amounted to approximately $110 billion globally, and approximately 31.82% of the total losses ($35 billion) were borne by insurance companies.

However, this proportion is much lower in China. For example, the 2008 Chinese winter snowstorms caused by low temperatures led to direct economic losses amounting to 151.65 billion RMB, of which insurance company payments covered only 6.039 billion RMB, for a relief ratio of only 3.98%. Similarly, the 2008 Wenchuan earthquake caused 845.1 billion RMB in direct economic losses, and insurance payments amounted to only 713 million RMB, for a relief ratio of only 0.08% (Table 5). The 2013 Ya An Earthquake caused approximately 10 billion RMB in direct economic losses, but insurance payments amounted to only 14.30 million RMB, for a relief ratio of approximately 0.14%.

2.3. *Social and charitable contributions*

Chinese people honour the traditional virtue of helping each other. In the past, all sectors of society have made charitable contributions after major natural disasters to support relief

Table 5. China's natural disaster insurance payments.

	Insurance payments (100 million RMB)	Direct economic losses (100 million RMB)	Relief ratio
Snow and ice storms	60.39	1516.5	3.98%
Wenchuan earthquake	7.13	8451	0.08%
Ya An earthquake	0.143	100	0.14%

Note: The yearbook does not directly provide the insurance payments for the three cases of natural disasters; the data come in the form of calculations from disclosures of information provided by various insurance companies in the yearbook.
Source: 'China Insurance Yearbook'.

efforts; thus, charitable contributions have become another important source of disaster relief funds.

Table 6 lists the total amount of charitable donations and expenditures for disaster relief from 2007 to 2016. To demonstrate the significance of social and charitable contributions to the disaster relief system, Table 6 also shows the total losses caused by natural disasters in the relevant years for comparison purposes. Over the past ten years, on average, 16.98% of all charitable contributions in China went towards disaster relief. However, the social charitable contributions relief ratio is 2.53%, which is still below the government's average fiscal disaster relief ratio of 3.98%.

2.4. Assistance from social organisations

According to the statistics of the Ministry of Civil Affairs, as of the end of 2015, there were 662,000 social organisations in China (Statistical Communique of Social Service Development, 2015). After the disaster, some social organisations stopped their routine programmes and turned to rescue operations. According to China's current social organisation registration management system, the teams and volunteer organisations that provide rescue services are mostly social groups and social service organisations, such as the Blue Sky Rescue Team and the Outdoor Sports Rescue Team. There are also some foundations that focus on disaster relief. Although the number is relatively small, they are playing an important role in organisational mobilisation, such as the China Red

Table 6. Social and charitable donations and disaster relief 2007–2016.

Year	Sum of charitable donations (100 million)	Relief expenses from donations (100 million)	Chari. Donation ratio of relief	Natural disaster total losses (100 million)	Relief ratio
2007	309.25	37.11	12.00%	2363.0	1.57%
2008	1070	769.59	71.92%	11,752.4	6.55%
2009	332.78	22.99	6.91%	2523.7	0.91%
2010	601.7	113.44	18.85%	5339.9	2.12%
2011	845.00	104.30	12.34%	3096.4	3.37%
2012	700.00	132.60	18.94%	4185.5	3.17%
2013	989.42	127.93	12.93%	5808.4	2.20
2014	1046.26	56.08	5.36%	3373.8	1.66
2015	1108.54	64.74	5.84%	2704.1	2.39
2016	1392.94	65.47	4.70%	5032.9	1.30
Average	839.589	149.43	16.98%	4618.01	2.53

Note: The first official report on charitable donations covered 2007 and was released by the Ministry of Civil Affairs' charitable Coordination Office in 2008.
Source: 'China Charitable Contributions Report' (2008–2016) and 'Social Services Development Statistical Report' (2010–2016).

Cross Foundation (Xing, 2017). In addition, some international NGOs are involved in disaster relief.

Apart from financial assistance, as discussed in 2.4, social organisations can participate in disaster relief work in several forms, such as daily disaster prevention and mitigation education, accelerating the speed of information dissemination (through internet, etc.) and helping to improve the information disclosure system.

2.5. Self-help and mutual aid between victims

Historically, Chinese people have had to rely on their own savings for relief from disasters, supplemented by borrowing, contributions and other sources of funding. Currently, victims' self-help and mutual assistance are the main form of post-disaster relief.

Taken together, the government's fiscal relief, charitable contributions and commercial insurance payments account for a rather small share of disaster losses; hence, by far the largest proportion of economic losses after disasters is borne by the victims.

3. Critical discussion of the disaster relief system in China

3.1. Government assistance

Although government assistance is the most important official channel for disaster relief in many developing countries, there are some clear limitations related to relying so heavily on the government.

First, it is difficult to effectively optimise government assistance. Current government assistance funds are generally based on past experiences with disaster losses and are allocated based on the Ministry of Finance's annual revenue and expenditure plans. According to current Chinese law, special funds for disaster assistance can only be used in the year the funds are collected, so the remainder cannot be carried forward (Chen, 2010). Therefore, because government budgets are generally completed early in the year, if a year ends up with many disasters and heavy losses, funds will be lacking, whereas in years with fewer disasters, there will be an over-allocation of relief funds. Moreover, government assistance is mainly concentrated on post-disaster relief; thus, it usually lags behind the events, which is not conductive to disaster prevention.

Second, reliance on the government reduces the efficiency of indemnity. Government assistance is derived from public fiscal expenditures, the focus of which is social justice, but the efficiency of such compensation is not considered. Meanwhile, it is also difficult for the government to effectively supervise the implementation of disaster relief.

Third, long-term government assistance can lead to psychological dependence on the government by victims in disaster-prone areas. This might cause people to not utilise insurance or take proactive measures to prevent disaster losses. Bishawjit, Khan, and Joachim (2011) researched the cyclone SIDR 2007 in coastal Bangladesh by random sampling and found 'dependency on relief works' among affected communities waiting for relief and reconstruction materials, which made them more 'vulnerable' to other calamities. In the long run, such dependency increases the poverty ratio and keeps the communities in a 'vulnerability trap' with regard to any type of disaster.

3.2. Insurance

Insurance is a way of managing risk, as concentrating and transferring risk helps to reduce uncertainty for people facing a disaster. Insurance can enhance a community's overall ability to resist risks and promote economic development. Disaster insurance for individuals, businesses and governments provides these actors with funds in the event of disasters and thus helps them recover. Without disaster insurance, social development might stall and stagnate (Swiss Re, 2014). Because of their ability to provide economic compensation and effectively transfer risk, insurance payments are increasingly becoming the most important channel for disaster relief internationally.

As an institution, commercial insurance is embodied in insurance contracts with the precondition of premium payments by the insured to fulfil the contractual obligation, and insurance payments are made in accordance with the principle price paid for claims, which makes such payments essentially different from government assistance and charitable contributions. The risk can also be diversified and transferred through reinsurance, institutional arrangements or catastrophe securitisation, thereby increasing the efficiency and effectiveness of commercial insurance payment relief channels (Zhao, 2009).

Unlike commercial insurance, a catastrophe insurance system includes individuals, social groups, domestic and foreign insurance companies, special organisations, the entire insurance market, and government and international organisations. Risks are shared through a hierarchical system. In the case of catastrophe insurance, the first tier is insurance payments by commercial insurance companies, while the second tier includes reinsurance companies and special institutions that assume the corresponding insurance responsibility based on contracts with the original insurer. Then, the government chooses whether to transfer the insurance risk to the entire market, as it is the ultimate 'insurer' that provides the excess portion of relief and a certain degree of security and loans to offset the companies' operating risk.

In terms of the social management of insurance, commercial insurance fosters social management innovation because insurance companies become actively involved in underwriting risk through scientific means of disaster prevention in advance. In addition to conducting risk assessments on the insured, insurance companies help policyholders actively carry out preventative measures before disasters occur (Wu, 2008). After disasters, insurance companies survey the scene for claims and use disaster prevention and mitigation technologies in a timely manner to reduce the scope of losses to the lowest level possible. They then pay the necessary compensation to the insured and help policyholders carry out reconstruction and quickly alleviate the negative impacts of disasters.

Catastrophe insurance is relatively more complicated. The social management of insurance is basically the same as the abovementioned commercial insurance. However, non-market catastrophe insurance is often more risk-oriented due to its greater risk of adverse selection, such that the publicity of risk management-related knowledge is more highlighted. Because of the government's participation and guidance, the intensity of publicity and education is greater, which can help strengthen the willingness and ability of the insured to prevent disasters, hasten recovery and reconstruction work in the affected areas, reduce the impacts on society, and maintain social stability.

3.3. Social and charitable contributions

Social and charitable contributions are unidirectional; that is, one party contributes a unit, and the other party has no obligation to accept the unit, such that the two sides do not share the same constraints, and charitable donors have no obligation to fulfil their commitment to relief. The funds raised from charitable contributions after the Wenchuan earthquake and the Zhouqu mudslides were very impressive. However, unlike government-channelled relief funds, these funds were short term and subject to restrictions imposed on the use of charitable donations. Therefore, charitable funding was not as efficient as funding from the government. Moreover, charitable funding also lacks a reasonable planning and coordination arrangement for the efficient use of these limited funds. Overall, social and charitable contributions cannot be relied on as the main source of disaster relief; they can act only as a necessary complement to a more institutionalised disaster relief system.

Moreover, the information asymmetry among different social charitable donations and disaster relief affects the efficiency of disaster relief (Yang, 2017). In the Lushan earthquake, 11.5% of social organisations never contacted government agencies to acquire the necessary information (Yang, 2017), and the three main reasons for not getting in touch with related government agencies were 'no consideration, no need, no channels' (Wang & Zhu, 2015). In this case, donors would only be able to obtain relevant information in the disaster area through the media or other channels. Without accurate information, the donors would not know how much donations should be contributed, so that the donations might be either too little to cover the disaster losses or too much, thereby wasting social resources. On the other hand, disasters with less damage are often not known to donors, which causes asymmetry in the proportion of donation funds and disaster losses.

3.4. Self-help, mutual aid and social organisation assistance

Victims helping themselves and assisting each other are an inevitable part of disaster relief and are becoming increasingly important nowadays, as the China Natural Disaster Ordination has for the first time cited 'self-help' as one of its disaster relief principles. Scholarship has shown that informal relations, particularly neighbours, regularly serve as actual first responders (Aldrich, 2012; Horwich, 2000; Shaw & Goda, 2004). It is true that self-help and mutual aid are highly important in disaster relief; self-help and mutual assistance capabilities should be enhanced (Deng, Su, Gao, & Sun, 2017; Guo et al., 2018). However, in the view of economic compensation, because disasters are bound to be devastating to individual and collective wealth, the speed and size of wealth accumulation by this kind of disaster relief are restricted, which results in limited emergency relief funds. From a risk management perspective, victims' self-help and mutual aid are very inefficient responses to disasters. The risk in self-help and mutual aid is retained and not transferred or reduced, so that adequate contingency funds are required to compensate the loss of uncertainty. However inconsistently, people in some affected areas are lacking risk awareness and capital strength, and most of them will not establish long-term risk prevention awareness for an extraordinary disaster (Su, Zhao, & Tan, 2015).

In addition, victims' funds for self-help and mutual assistance must be accumulated over a period of time. If a disaster causes substantial losses in this period, this channel of relief can be blocked, as savings and reciprocal arrangements can be ineffective for covariant, catastrophic risks that impact savings and credit institutions, as well as families and neighbours (Linnerooth-Bayer & Hochrainer-Stigler, 2015). Additionally, if the victims accumulate a sum of self-rescue funds over a long period, the management and use of these funds becomes a problem because leaving such funds idle affects their value. Moreover, such funds are generally smaller, so liquidity requirements are high, which also limits the choice of investment portfolio.

With reference to assistance from social organisations in disaster relief, although they have the advantages of rapid response, extensive service areas, and meticulous work, there are problems related to legality, participation, cooperation, resources, and ability, which restrict their ability to participate. This is mainly due to the imperfection of the legal system, the defects of the management system, the lack of preferential incentive mechanisms, the lack of benign interaction and cooperation mechanisms, and the defects in the NGO's own system (Meng, 2014).

4. Suggestions for reconstructing China's disaster relief system

4.1. Improving the efficiency of government assistance

As discussed above, the government prepares disaster aid through its fiscal budget every year, and this capital is used for disaster relief and post-disaster reconstruction, actually, according to the natural disaster relief regulation, once a disaster happens, the governments use appropriation to buy relief materials and services, subsidising people to reconstruct their houses, and even comfort the family members of the victims, which can be categorised as passive spending. However, government assistance in response to disasters should be geared towards the prediction, prevention and mitigation of disasters through research and construction more than it currently is (Zhang & Wang, 2013). Governmental fiscal expenditures related to disaster prevention are rather weak; for instance, the United States has a risk distribution map for various natural disasters over the last century, and in this respect China is currently in an exploration stage, although the Chinese Academy of Sciences has established a risk map for debris flow disasters (Bao, 2011). Moreover, with regard to government assistance, there are still different ways to prevent disasters, such as regional risk assessment reports issued by official bodies.

The government can also use assistance funds to buy insurance products to transfer the catastrophe risk and reduce the fiscal burden for disaster relief. For example, Mexico's government has established a Natural Disaster Fund (FONDEN) to safeguard the country's financial stability while ensuring that relief funds and insurance payments are mainly used to provide disaster relief funding for infrastructure repair and insurance coverage for those who are unable to insure themselves from disasters (Swiss Re, 2009). Thus, the government used national financing to buy insurance products from Swiss Reinsurance. In this case, the Mexico government takes on the role of the insured and works with insurance companies to manage disasters in a complementary way, thus using insurance functions and adopting insurance company management models to achieve effective disaster management (Xu, 2011b). According to the government of Mexico's insurance purchase agreement, when Mexico has an earthquake measuring 6.5 or above on the Richter scale, insurance companies that have contracts with the Mexican government will provide a payout of $450 million in insurance indemnity, and these claims will be used for emergency aid after the earthquake. This model thus provides the Mexican government with reliable funds for relief in the event of a major earthquake.

4.2. Establishing a shared risk system

The suddenness and frequency of natural disasters create a need for a multifaceted and comprehensive relief system. Developed countries' experiences show that shared risk systems with the participation of national financing, insurance companies and the insured parties are an effective model for catastrophe risk management. China's government should improve its disaster risk management system by forming a risk-sharing mechanism and implementing regulations and a basic framework for insurance policies (Bao, 2011).

A shared risk setup should combine policy insurance and commercial insurance. The establishment of an operating mechanism for the insurance industry depends on individuals at risk decreasing and neutralising the level of risk. However, it is difficult to develop insurance products against catastrophes that have a small probability but substantial losses, as the consciousness of the people cannot be relied upon, which limits the development and supply of commercial insurance (Li, 2013). Therefore, many countries have adopted policy insurance measures to address the risk from national disasters. For example, the US federal government established a sound policy insurance system in high-risk and low-profit commercial insurance fields such as the National Flood Insurance Program (NFIP) and the Federal Crop Insurance Corporation (FCIC). The advantage of policy insurance is that it helps people face common risks under a unified insurance management policy so that the efficiency of risk management as well as insurance payments for victims are ensured. However, the NFIP in the United States depends on voluntary participation, which can lead to adverse selection and high premiums. If such a system is made mandatory, that is, the government transfers the risk to taxpayers, it may cause dissatisfaction among people facing low levels of risk (Wang, Wang, Fan, & Teng, 2008). Hence, risk-based premiums are needed, and such insurance contracts could be complemented by multiyear home improvement loans provided by the government or commercial banks to increase the participation ratio (Michel-Kerjan & Kunreuther, 2011).

4.3. Establishing a suitable disaster insurance mechanism

China's insurance market has developed rapidly in recent years. In 2015, the contribution of the global insurance market had increased to more than 30% (Zhao, 2017). The life insurance business has become more prominent, and property insurance has maintained uniform growth. However, the development of catastrophe insurance has been slow. China's insurance market has a disaster pay-out ratio of less than 5%, and the global average exceeds 30% (Li, 2011). China's domestic supply of disaster insurance is mainly in the area of property insurance. Most common disaster risks are covered, including typhoons, rainstorms, lightning strikes and mudslides. China's current catastrophe insurance pilot programme includes government purchases of catastrophe relief insurance for all citizens from commercial insurance companies in addition to a government-funded catastrophe insurance fund and commercial catastrophe insurance. This pilot programme covers a wide range and plays a fundamental role in security. In some areas, the earthquake insurance pilot project is led and paid for by the government but supplemented by commercial insurance, thus reflecting the government's desire to cover catastrophe risk through the joint insurance model.

The demand for this model stems from China's economic characteristics, as the country remains in the development stage and has an undeveloped insurance market. Through the joint insurance model, the government can not only reduce its financial burden but also prevent insurers from bearing excessive catastrophe risk, thus maintaining the stable development of the insurance industry. In this sense, China should accelerate the establishment of special legislation with regard to the main types of catastrophes to develop a legal system applicable to catastrophe insurance. In addition, coercive measures should be adopted to improve the insured rate among residents. The government and the market share important responsibilities in the joint insurance model, but these obligations must be clearly defined. Specifically, the government should be primarily responsible for the improvement of public goods such as disaster prevention and mitigation, tax legislation that promotes the supply of products, and laws that require residents to be insured. In contrast, the commercial insurance market should act as risk management experts through the use of professional technology and innovative insurance products. In addition, China's vast territory requires that the catastrophe insurance system be established with full consideration of central, local and community relations. The establishment of local programmes should be based on local catastrophe situations and regulations, and the local programmes should obtain the approval and support of the central authorities. Community-based disaster reduction activities should be promoted. The gradual establishment of a premium income catastrophe fund (with concessional loans from the government) can facilitate the development of a multi-level risk-sharing mechanism.

Moreover, the management and cooperation between the Emergency Management Ministry and central, local and community entities also need to be improved. Other than China's earthquake catastrophe insurance community, which was established in 2015, there is no special institution for catastrophe insurance. The insurance industry is supervised by the China Insurance Regulatory Commission, and the provinces and cities also have supervisory branches. To ensure the full implementation of the government's role in the catastrophe insurance system, the relations between the central Insurance Regulatory Commission and local insurance regulatory commissions should be improved as well as the relations between the central government, local governments and communities. First, in the case of earthquake and flood disasters and in accordance with the scope of the specific disasters, the Insurance Regulatory Commission can jointly establish plans with local governments to improve both the level of efficiency and local conditions. Second, the central, local and community governments should work from the bottom up, with local governments developing disaster mitigation projects for area-specific disasters and seeking financial help from the central level. From the top down, the central government can enact national disaster laws tailored to specific disasters and urge local governments to implement disaster mitigation projects. Local governments should take lessons from the experience of the NFIP in the USA by engaging in disaster reduction activities and improving the overall participation in catastrophe insurance.

4.4. Enhancing the role of commercial insurance

The role of commercial insurance in China's current disaster relief system is negligible. For example, in 2008, insurance payments represented a ratio of only 0.59% of disaster losses, which is far below the international average of 36% (Xu, 2011a). In 2013, Heilongjiang Province was affected by floods that caused direct economic losses of 19.1 billion RMB, including 6.31 billion RMB in agriculture losses. The estimated insurance payments reached more than 2.9 billion RMB, which was over 15% of the total disaster losses (Guan, 2014). Although the Chinese insurance system has made great progress in the past few years, it is still far from the world average. The low rate of disaster relief from commercial insurance payments is caused by a variety of factors. First, it is difficult for insurance providers to determine rates due to a lack of relevant historical data. Second, cumulative and concentrated payouts for catastrophe losses pose great challenges for insurance companies' underwriting capacity. Finally, most Chinese people have a poor sense of the need for catastrophe insurance, which leads to a low level of demand for insurance and high insurance fees. All of these factors limit the present role of commercial insurance.

China should learn from developed countries and increase the contribution of commercial insurance in addressing catastrophe risk. The government should use insurance schemes, tax incentives and shared compensation to promote the development of commercial insurance. Because catastrophe insurance often increases the solvency risk for insurance companies, sometimes even leading to bankruptcy with serious disasters, the government should support insurance companies in providing compensation (Swiss Re, 2010).

4.5. Improving people's risk consciousness

Changing the disaster relief system also requires changing people's approach to risk management from passive to proactive disaster prevention. Affected people were responsible for 83.26% of the economic losses from disasters in 2008 (Shi & Xu, 2011), a burden that has seriously affected the living conditions of the disaster victims. In addition, although Chinese people generally have a correct perception of the hazards in the areas they live, the perceived risk does not directly affect disaster insurance participation because the people who face a higher disaster risk may not necessarily recognise the importance of insurance as a coping mechanism (Wang et al., 2012). Therefore, to effectively reduce the proportion borne by the affected population and increase the participation rate, the risk of possible losses should be reduced through housing design and construction as well as localised science. In addition, affected populations should be encouraged to purchase insurance products to transfer the risk and improve the efficiency and scale of victims' self-help and mutual assistance through collective risk self-help funds. On the other hand, as shown in the survey conducted by Wang et al. (2012), those who live in regions with a higher integrated hazard index score tend to expect the government to take the responsibility for disaster losses, which was the most important factor preventing them from purchasing disaster insurance. Thus, to strengthen the capacity for disaster prevention and allow the government to increase its investment in related equipment and infrastructure, the government should actively publicise disaster insurance and the ability of insurance companies to help people make risk management and insurance plans. Thus, the government and insurance companies can together improve the efficiency of addressing claims and educate people about the importance of the role of insurance in the disaster relief system.

5. Conclusions

Although government assistance, charitable donations and commercial insurance play important roles in China's disaster prevention and reconstruction setup, even when combined, they still account for a very low ratio of the total losses. Moreover, as described above, the current disaster relief system suffers from a number of problems, such as the low efficiency and passiveness of the government, the uncertainty related to charitable donations, and the limited role of commercial insurance compared to developed countries.

It is imperative to reform and restructure China's existing disaster relief system to improve its effectiveness and efficiency. The corresponding measures that should be taken include (1) active disaster prevention activities to improve the efficiency of government assistance; (2) a system of risk sharing between the government, commercial insurance companies and victims; (3) an increased role for commercial insurance; (4) improved risk consciousness among the people for self-help during disasters; and (5) an increased participation ratio for disaster insurance. All of China must work together to achieve the effective functioning and sustainable development of the country's disaster relief system. This paper has analysed and provided suggestions for China's disaster relief system at the macro level. More detailed and specific measures should be discussed in future research.

Notes

1. The original 'China Insurance Regulatory Commission' and the 'China Banking Regulatory Commission' have merged into the 'China Banking and Insurance Regulatory Commission' in 2018.
2. Data source: the local branches of CBIRC.
3. A mismatching case is one in which, for example, rural housing insurance only covers the rural houses but excludes other properties (Li et al., 2015). Another example is that those people who face higher disaster risks may not necessarily recognise the importance of insurance as a coping mechanism (Wang et al., 2012).
4. Yunnan and Sichuan are both provinces where earthquakes are common, so that a targeted earthquake insurance has been offered. As the losses caused by an earthquake can usually be devastating, more insurance companies need to work together to share the risks.
5. In 1969–1998, catastrophe losses ranked 3rd among the reasons for the bankruptcy of American insurance companies (Hu, 2008).

Disclosure statement

No potential conflict of interest was reported by the author.

Funding

This work was supported in part by National Natural Science Foundation of China [grant number 71673050] and by Gaofeng Research Project of School of Economics, Fudan University.

References

Aldrich, D. (2012). The power of people: Social capital's role in recovery from the 1995 Kobe earthquake. *Natural Hazards, 56*, 595–611.

Bao, W. (2011). International comparison of disaster insurance. *Exploration and Free Views, 6*, 114–117.

Bishawjit, M., Khan, R., & Joachim, V. (2011). Social vulnerability analysis for sustainable disaster mitigation planning in coastal Bangladesh. *Disaster Prevention and Management: An International Journal, 20*(3), 220–237.

Chen, S. (2010). *Flood insurance study based on an analysis of supply and demand*. Beijing: China Yanshi Press.

Deng, Y., Su, G., Gao, N., & Sun, L. (2017). Investigation and analysis of the importance awareness of the factors affecting the earthquake emergency and rescue in different areas: A case study of Yunnan and Jiangsu Provinces. *International Journal of Disaster Risk Reduction, 25*, 163–172.

Fan, H. (2016). Research on the housing earthquake insurance models. *Shanghai Insurance, 2*, 45–48.

Guan, J. (2014). The risk prevention ability in Hei Longjiang Provincce of agricultural insurance is insufficient. *Economic Information Daily*. Retrieved from http://jjckb.xinhuanet.com/2014-07/02/content_510983.htm

Guo, L., He, B., Ma, M., Chang, Q., Li, Q., Zhang, K., & Hong, Y. (2018). A comprehensive flash flood defense system in China: Overview, achievements, and outlook. *Natural Hazards, 92*(2), 727–740.

He, L. (2017). Study on interlinkage between farmhouse insurance and catastrophe insurance in China. *Journal of Northwest A&F University (Social Science Edition), 17*(4), 121–127.

Horwich, G. (2000). Economic lessons of the Kobe earthquake. *Economic Development and Cultural Change, 48*, 521–542.

Hu, Y. (2008). Looking at the policy relief from the bankruptcy of overseas insurance companies. *Public Financial Advisor, 2*, 84–85.

Li, S. (2011). A gap in China's disaster compensation and the establishment of the catastrophe insurance system need to speed up. *China Youth Daily*. Retrieved from http://finance.people.com.cn/insurance/GB/14133369.html

Li, X. (2013). Research on catastrophe risk management model in China. *Times Finance, 3*, 228–238.

Li, C., Zhang, N., & Guo, L. (2015). Flood insurance: Demand, supply and public policy. *Insurance Studies, 5*, 51–59.

Linnerooth-Bayer, J., & Hochrainer-Stigler, S. (2015). Financial instruments for disaster risk management and climate change adaptation. *Climatic Change, 133*(1), 85–100.

Liu, J. (2009). Revelations on disaster economics in China's economic development. *Journal of Hunan Administration Institute, 55*(1), 71–72.

Meng, T. (2014). Interpretation and system reconstruction of non-governmental organizations participating in disaster relief—taking Wenchuan earthquake as an example. *Journal of Southwest Minzu University (Humanities and Social Science), 35*(02), 87–91.

Michel-Kerjan, E., & Kunreuther, H. (2011). Redesigning flood insurance. *Science, 333*(6041), 408–409.

Ministry of Civil Affairs of China. (2001–2017). *China civil affairs statistical yearbook*. Beijing: China Statistics Press.

Ministry of Civil Affairs of China. (2008–2016). *China charitable donations report*. Beijing: China Society Press.

National Disaster Reduction Committee. (2016, March 24). National emergency plan for natural disaster relief.

National Emergency Plan for Natural Disaster Relief. (2016). Retrived from http://www.gov.cn/yjgl/2006-01/11/content_153952.htm.

Shaw, R., & Goda, K. (2004). From disaster to sustainable civil society: The Kobe experience. *Disasters, 28*, 16–40.

Shi, L., & Xu, X. (2011, April 15). Promote multi-participation and improve the efficiency of disaster relief. *China Insurance News*. p. 2.

Statistical Communique of Social Service Development. 2015. Retrived from http://www.gov.cn/shuju/2016-07/12/content_5090289.htm

Su, Y., Zhao, F., & Tan, L. (2015). Whether a large disaster could change public concern and risk perception: A case study of the 7/21 extraordinary rainstorm disaster in Beijing in 2012. *Natural Hazards, 78*(1), 555–567.

Swiss Re (Ed.). (2009). *Natural catastrophes and man-made disasters in 2008*. Zurich: Swiss Re Press, p. 14.

Swiss Re (Ed.). (2010). *Natural catastrophes and man-made disasters in 2009*. Zurich: Swiss Re Press, pp. 13–14.

Swiss Re. (2014). *Natural catastrophes and man-made disasters in 2013*. Zurich: Swiss Re Press. p. 4.

Swiss Re (Ed.). (2017). *Natural catastrophes and man-made disasters in 2016*. Zurich: Swiss Re Press, p. 1.

Wang, J. (2008). Improving disaster relief needs to develop catastrophe insurance. *China Disaster Reduction, 5*, 2.

Wang, M., Liao, C., Yang, S., Zhao, W., Liu, M., & Shi, P. (2012). Are people willing to buy natural disaster insurance in China? Risk awareness, insurance acceptance, and willingness to pay. *Risk Analysis, 32* (10), 1717–1740.

Wang, Y., Wang, H., Fan, G., & Teng, Y. (2008). The enlightenment of the disaster loss sharing mechanism in the United States and Japan——based on insurance perspectives. *Insurance Research, 6*, 19–22.

Wang, D., & Zhu, M. (2015). A survey of status quo with respect to Chinese social organizations participation in disaster relief – taking the earthquake relief of Lushan earthquake as an example. *China Emergency Management, 106*(10), 20–23.

Wei, H., & Zhang, S. (2012). The "difficulties" and breakthroughs of government intervention in the catastrophe insurance business model. *Insurance Research, 1*, 21–29.

Wu, X. (2008). The role of insurance industry in disaster management. China *Democratic League Disaster and Social Management Expert Forum.*

Xing, Y. (2017). The realization mechanism of social organizations participating in disaster relief in China from the perspective of collaborative governance. *Administration Reform, 08*, 59–63.

Xu, X. (2011a, August 12). Change in disaster environment and the new mission of China's insurance industry. *China Insurance News*, p. 2.

Xu, X. (2011b). The international model of insurance industry participating in disaster management. *China Finance, 9*, 66–67.

Yang, T. (2017). Research on non-profit organizations participate in government emergency management —A case from earthquake relief of the Lushan earthquake. *Advances in Social Sciences, 6*(6), 716–722.

Zhang, J., & Wang, H. (2013). Establishing a national comprehensive disaster risk management mechanism. *China Disaster Reduction, 3*, 36–37.

Zhao, Y. (2009). Catastrophe insurance system architecture analysis and catastrophe insurance system in China. *Finance and Trade Economics, 9*, 70–76.

Index